THE UNBOUNDED MIND

Breaking the Chains
of Traditional
Business Thinking

IAN I. MITROFF

AND

HAROLD A. LINSTONE

New York   Oxford
OXFORD UNIVERSITY PRESS
1993

Oxford University Press

Oxford   New York   Toronto
Delhi   Bombay   Calcutta   Madras   Karachi
Kuala Lumpur   Singapore   Hong Kong   Tokyo
Nairobi   Dar es Salaam   Cape Town
Melbourne   Auckland   Madrid

and associated companies in
Berlin   Ibadan

Library of Congress Cataloging-in-Publication Data
Mitroff, Ian I.
The unbounded mind : breaking the chains
of traditional business thinking /
by Ian I. Mitroff and Harold A. Linstone.
p.   cm.   Includes bibliographical references and index.
ISBN 0-19-507783-0
1. Business education—United States.
2. Critical thinking—Study and teaching (Higher)—United States.
I. Linstone, Harold A.
II. Title.   HF1131.M58   1993   650'.071'173—dc20   92-15708

2 4 6 8 9 7 5 3 1

Printed in the United States of America
on acid-free paper

For C. West Churchman
teacher, friend, and colleague

Preface

The need for new thinking is like the weather. Everyone talks about it, but beyond that, no one does much about it. This book not only talks about the need for new thinking, but attempts to show precisely why and how it can be accomplished.

It is precisely the inability of old ideas to meet the challenges of our age that makes new thinking more critical than ever. If ours is indeed the "new knowledge/information age," then ideas are obviously at its core. The need for a philosophy of and for this age is more desperate than ever.

Professional Education Today

The authors feel strongly that professional education in the United States today is seriously deficient. Graduates emerge with a variety of specialized tools of analysis that ill equip them for the complex real-world problems they encounter in their professional lives. Our perspective is business school, the professional schools we know best, but we also believe that the problem is common to other professional schools, such as engineering and law.

While this is not primarily a book on education or curriculum reform, it does contain a blueprint and a prescription for the kinds of thinking skills and knowledge that students, workers, and executives at all levels will need to master for life now and into the twenty-first century. If America is to regain its economic standing, then it will have to turn out students and executives at all levels who can identify outdated assumptions, and then critique, challenge, and replace them with new ones.

Philosophy in Action

Although, from one perspective, this is a book of and on philosophy, it is philosophy with a new twist. We have attempted to make difficult ideas accessible by using academic jargon as little as possible, and, conversely, everyday examples as much as possible.

The book is an exercise in "applied philosophy" or "philosophy in action." For many years, the authors have taught a number of courses in "critical thinking," "creative problem solving," "brainstorming," "systems analysis, etc." to corporate executives, MBAs, and graduate students. Few of them had a formal background in philosophy. It was perhaps just as well they didn't, for it spared us from the awful task of "unlearning" a number of misguided notions about philosophy. Philosophy is neither primarily for, nor the exclusive province of professional, academic philosophers. To exaggerate if only slightly, with a few significant exceptions, professional philosophers have demonstrated a remarkable inability and incredible disdain for getting involved with practical problems. For this reason, we are especially critical of the role that academic philosophers—if not academics in general—have played in shedding light on vital issues.

It is one thing, however, to say that we are especially critical of the traditional role that academic *philosophers* and academicians have played in general. It is quite another to dismiss the fundamental role that *philosophy* has in illuminating important problems.

In today's world, economic success demands that one be able to examine problems from multiple perspectives. It also demands that one formulate multiple and conflicting definitions of critical problems. It further demands that one be able to examine how a problem will affect and be affected by different persons scattered throughout every level of society, if not the entire world. It necessitates the ability to treat two of the most critical aspects of all problems—their aesthetic and ethical dimensions. Everyone must learn these skills and more if we are to survive and prosper in the new world that is literally being created under us. How to accomplish this is the subject of this book.

The knowledge and examination of key assumptions are the main components of critical thinking. One of the best ways (but not the only) to engage in critical thinking is to examine the philosophical ideas and assumptions that form the base for the tools and techniques of professional education. In this sense, this is not a book on philosophy for the sake of philosophy. It is fundamentally an assessment of the adequacy of the foundations on which all professional education rests.

Los Angeles   I.I.M.
Lake Oswego, Ore.   H.A.L.
April 1992

Contents

THE UNBOUNDED MIND

Control of knowledge is the crux of tomorrow's worldwide struggle for power in every human institution.

Alvin Toffler,
Powershift
(New York: Bantam, 1990), p. 20

. . . Even if America were to devote more resources to education, simply more of the same would not prepare its youth for roles in flexible-system enterprises. At best, the current system of education prepares young people for preexisting jobs in high-volume, standardized production. Some students are sorted into professional ranks and trained in the manipulation of abstract symbols. Others are prepared for lower-level routine tasks in production or sales. Few students are taught how to work collaboratively to solve novel real-world problems—the essence of flexible system production.

Robert B. Reich,
The Next American Frontier
(New York: *Times Books*, 1983), p. 215

CHAPTER

1

The World That Was
and Is No More

In the past ten years, U.S. businesses have been challenged more seriously than in any previous period. This challenge is a direct response to the growing globalization of the world's economy—as large and as powerful as the U.S. economy is, it is now more affected by the economies of other nations than ever before. Consequently, the context in which U.S. business now operates has changed so dramatically that it is forcing a radical reassessment and redesign of almost every aspect of the modern factory and corporation.

It is no longer "business as usual."[1] The worrisome trade deficits that the United States has accumulated as a result of the heightened competitiveness of foreign goods are a clear signal that for all practical purposes (very practical indeed), business today is global. There are no such things as local, protected markets. All markets are vulnerable to increased competition because of the often superior quality of goods produced by foreign manufacturers. No industry or sector of the economy is free from significant foreign competition or its future prospect.

In addition to the natural competition between individual companies, other factors not only make today's competition more intense but render it very different from previous eras. Economic competition occurs not just between individual firms but between nations: "government-supported" companies and industries. The result is no less than a worldwide competition or a *large-scale social experiment* between companies, industries, and entire governments regarding the de-

[1]I. I. Mitroff, *Business Not as Usual, Rethinking Our Individual, Corporate, and Industrial Strategies for Global Competition* (San Francisco: Jossey-Bass, 1987).

3

sign principles that are appropriate for conducting business into the next century.

The Factory as an Idea System

More than ever, the modern factory and business corporation are Idea Systems. They not only produce goods, but because they are so dependent on ideas as well, they serve as systems for the production and testing of ideas. Thus, the modern factory and corporation cannot only be examined from the standpoint of input and flow-through of materials to final end products, but also viewed from the initial input and flow-through of ideas. If the modern factory and business corporation are in effect organizational and social experiments for testing new ideas crucial to the production of quality goods that can compete worldwide, then we must examine in as systematic and comprehensive a fashion as possible the basis of these ideas. This examination is a central task of this book.

The New Service Factory: Have Factory, Will Travel

The emerging concept for the service factory[2] is one of many examples that illustrate how the globalization of businesses is revising the principles for doing business today. Because of increased competitiveness brought about by the global economy, products must directly meet the needs of consumers the first time out. *Every factory, no matter what it makes, must have a strong and integral service mentality. Every employee, no matter what his or her formal job title and responsibility, is in the service business.* For this reason, consumers are increasingly being invited to participate directly in the design of products, sometimes right on the factory floor itself.

Another aspect of the service factory is that factories themselves are moving beyond the notion of four fixed walls. In many cases, factories are literally traveling directly to consumers to build what they need. Thus, manufacturing can now be done either at a fixed site, in transit, or at the customer's location.

The Service Factory rejects the idea that factories must have four walls that are rigidly bounded in space and time. At the heart of this

---

[2]R. B. Chase, and W. J. Erikson, "The Service Factory," *The Academy of Management Executive*, 2(3), 191–196 (1988).

concept is the notion that a factory is an *unbounded, open system.* In other words, a factory must be extremely aware of and sensitive to its environment so that it can incorporate quickly developing information on the shifting needs and wants of consumers. As a result, nearly every one of the key and sacred assumptions that governed the design of business institutions in the past is being rethought and in many cases overthrown and replaced with fundamentally new assumptions.

A Brief Summary of the U.S. Industrial System Between World War II and the Present

To understand why a radical and thorough redesign of American business is needed, we must first realize how the total context in which business operates has changed. The starting point for our examination is the conditions found in the United States immediately after World War II.[3]

At that time, the United States was the richest, strongest, biggest, and most productive economy in the world. Its factories were the most modern. They contained the latest manufacturing technologies. Its work force was the best educated and most productive. Its factories produced the most advanced and highest quality goods. As a result, U.S. goods were in high demand. In return for their high productivity and the high demand for their goods, U.S. workers enjoyed the highest wages, benefits, and standard of living.

All of this was made possible by several distinct advantages the United States enjoyed over almost all other economies and nations. First, its resource base was incredible; the United States was blessed like no other nation in the abundance and extent of its raw materials. With very few exceptions, the United States was nearly self-sufficient—it rarely had to depend on the raw materials of others. Second, its internal domestic markets were so big, so broad, and, relatively speaking, so unsaturated for goods and services of all kinds, especially after the war deprivation years, that for all practical purposes U.S. manufacturers and businesses could ignore without harm all other foreign markets. Thus they concentrated almost exclusively on markets they not only knew more intimately but could control more fully and directly.

Third, our general technology base was the latest and most advanced, thus making our entire society the most modern for the times.

---

[3]For an in-depth discussion of the conditions after World War II and what's so different about today's world, see the Background References listed at the end of the book.

Fourth, as a result, the United States had an almost monopolistic lock on the importation of whatever raw materials it lacked. Because other nations didn't have the same productive and manufacturing facilities, they had few alternatives but to ship their raw materials to the United States below prevailing world market prices in return for final, finished products that the United States sold back to them.

Fifth, the United States had one of the finest infrastructures. Its transportation, communication, and educational systems were the best for the times. Sixth, consumers were not only starved for goods after the war, but their needs were relatively standardized and homogeneous. This meant that manufacturers had only to produce a limited number of distinct types and styles of goods. Whatever they produced would be gobbled up instantly, no questions asked. Seventh, consumer tastes were also rather predictable, stable, and long lasting. For example, forty years ago, the life cycle of a typical consumer product, say a refrigerator, was normally thirty years. This meant that it took thirty years for a refrigerator to diffuse its way through the entire population before meeting its "death" in the marketplace. A new product would first be bought by early adopters, for example, the avant-garde, the opinion leaders, the wealthy, the experimental, and so on. Then, as costs came down, it would slowly reach the rest of the population until its adoption was nearly complete, except for induced demand through yearly model changes. Thus, it was relatively easy to justify the huge costs involved in building big plants and equipment necessary to produce goods in large quantities for servicing mass markets. One could be relatively certain that the demand for goods would be there when they rolled off assembly lines into American stores.

Eighth, the cost of transporting raw materials, finished goods, services, information, and even knowledge itself between countries was relatively high. The two oceans acted as natural and, even more important, one-way buffers. Since we were stronger, and, so we thought, smarter than the rest of the world, we could penetrate markets anywhere if we desired. (For the most part, we were content not to.) But the huge costs involved prevented foreign manufacturers from setting up factories and distribution and dealer networks on our home shore. The grand result was that in effect we could ignore the rest of the world, but they could not ignore us.

Ninth, bigness was pursued almost indefinitely, that is, without limit, and solely for its own sake. Indeed, bigness followed almost automatically from the other conditions or advantages we had. If the United States possessed a decisive advantage in brains, talent, Yankee ingenuity, a well-educated and plentiful labor force, a rich abundance of relatively cheap raw materials, and consumer tastes that were largely predictable, enduring, and stable, then manufacturing organizations—and even government agencies—could grow almost without bounds and lose no efficiency in the process. The plain fact of the matter was that for those times organizations could by growing bigger achieve greater efficiencies and economies of scale. And they did precisely this for a relatively long period. Who after all could argue against big government and big business when in close partnership they had just fought the largest, "most successful" war in history? Like the land itself, which was so big it seemed it could never fully be conquered, organizations pursued growth without limit. In a word, the United States became the biggest, most successful economic engine—machine—the world had ever seen. And the term "machine," metaphor that it is, was the perfect description.

From its very beginnings, America has always had a deep and lasting love affair with technology.[4] If progress was the ultimate shining light of American society, then technology was its guiding hand-servant. And what better conception of a stable, smoothly running world, all of whose inputs were under control (raw materials, labor, infrastructure, consumer tastes and demand) than a "machine"? And if machines that were small and in isolation ran smoothly, what was there to prevent the total machine from running even better if they were all hooked together and the larger they became?

Given the crushing advantages and overwhelming success the United States enjoyed, what could possibly derail the American dream machine? Or had there always been a weakness lurking somewhere within? The answer is not only a resounding "yes," but also a phenomenon we term *the failure of success.*

The Failure of Success

Ironically, it was not the initial failure of the American experiment itself that led to the subsequent difficulties the United States is currently experiencing. Instead, it was America's huge, unparalleled success. More precisely, it was our failure as a culture to understand that it was only a very special, limited set of conditions that caused America's temporary success, no matter how long it seemed to last. We confused and took for granted short-term, temporary, *conditions* as permanent

[4]J. Lears, *No Place of Grace, Antimodernism and the Transformation of American Culture 1880–1920* (New York: Pantheon, 1981).

advantages that existed for all time. We failed to understand and take the necessary corrective actions to prevent nearly every single one of our considerable advantages from turning into major disadvantages.

While we rested on our laurels after World War II, the other nations of the world, with their economic and infrastructures either destroyed or undeveloped, had nowhere to go but up. The United States itself even became one of the biggest contributing factors to the rebuilding of the other economies through the Marshall Plan and other programs.

What were once decisive advantages for the United States turned steadily over time into disadvantages. Unlike the United States, Japan had to rebuild its almost totally destroyed manufacturing base after World War II. Further, unlike the United States, which had always possessed an overabundance of cheap raw materials, Japan was constantly faced with the opposite, a general impoverishment of raw materials. Thus, when Japan recreated its manufacturing base, it not only equipped it with the next generation of even more modern and more efficient plants and facilities, but rebuilt it so that it would squeeze the most out of a limited raw material base. Even more astounding, Japan learned how to turn its meager raw material base, its disadvantages, into crushing advantages.

Unlike the United States, which had so many raw materials it could literally afford to squander them in industrial processes that in a few years after the war were becoming steadily more and more inefficient because of their bigness and advancing age, Japan had no choice but to opt for processes that conserved energy and materials in the most efficient manner possible. Such processes usually involved the construction of a larger number of smaller factories. The mini steel mills are just the latest developments in this long line of thinking and experimentation. But smallness, it turned out, also possessed other decisive advantages. By building smaller, more efficient plants, the Japanese, as well as the Koreans, and later the Taiwanese, were learning other invaluable lessons that would serve them extremely well some twenty-five years later. They were acquiring flexibility and interdependence. They were learning that on every dimension and every level of society "less can be more."

Since smaller plants were easier to design, maintain, and manage than big ones, they permitted much more flexible manufacturing processes capable of shifting quickly from one production mode to another. Thus, in the 1970s and 1980s, when consumer tastes not only became more differentiated and segmented but literally changed overnight, the Japanese learned how to adjust production runs to accommodate those tastes faster than Americans. Other nations soon learned to follow suit.

The success of the Benetton stores is only one of many dramatic examples. These stores are linked all over the world via computers; it has now become possible to orchestrate overnight style and color changes that would have boggled the mind a decade ago. To do this, Benetton's management has had to forge extremely close ties with clothing mills to produce at a favorable cost savings the latest fashions that are at the leading edge of what consumers want or can be made to want.

The ability to quickly shift production runs became an extremely important asset as the cost of transporting huge volumes of raw materials, information, and knowledge decreased enormously. The vast distances and geographical barriers between nations were no longer the natural buffers they had been for so long. This in turn meant that foreign manufacturers could invade the home turf of domestic manufacturers more rapidly and effectively. With speedier and cheaper ways of transporting huge quantities of information via the computer, foreign companies could communicate more easily with their offshore plants and facilities. It was also easier to develop nationwide transportation and dealer networks overseas. It would have been unthinkable forty years ago for new car lines such as Infiniti and Lexus to be established almost overnight.

The United States that had always relied on the impenetrability of its huge internal domestic markets was now at a considerable disadvantage. It knew only how to market to that one market. Its manufacturing and plant facilities were not only increasingly out of date, and hence growing inefficient, but they were becoming more inflexible as well. The result was that U.S. manufacturers could not even keep up with the constantly shifting tastes of their own consumers. Not only were they losing market share in world markets where they once had enjoyed commanding leads (e.g., in the 1950s the United States had nearly 50% of world steel markets, yet that share is now 30% and heading lower[5]; the Japanese now produce a larger total volume of cars than we do, a situation unimaginable just two short decades ago), but they were in danger of reducing significant shares of their own internal markets. By not learning how to market and compete globally as the Japanese had done out of necessity, the United States lost ground in its domestic as well as global markets.

If there was anything good that came out of World War II for the Japanese, it was that they had faced a total overhaul of their entire manufacturing apparatus. When one's infrastructure is completely de-

---

[5]I. C. Magaziner, and R. B. Reich, *Minding America's Business: The Decline and Rise of the American Economy* (New York: Vintage, 1983).

stroyed, the ground for accomplishing significant change is made infinitely easier. Crisis may be the best, if not the only, teacher on how to create an economy that is better matched to the needs of today's world.

Interconnectedness

Step by step, we have been led inevitably to the concept of interdependency. If Japanese firms were to reap the advantages of smallness (e.g., the ability to shift and adapt quickly to changing consumer tastes), then they had to coordinate the efforts of innumerable groups, or stakeholders, both inside and outside the boundaries of their organizations. For example, if the Japanese were to cut the tremendous costs of holding huge amounts of raw materials and finished products in inventory, then they had to reduce the size of their factories. But to do this they had to forge close ties with their external suppliers whom they could trust without question to deliver the right materials at the right time and in the right spots daily—in many cases hourly.

American factories have always been big holding pens of parts and finished products. In part, this occurred because the cost of buying and maintaining land had never been as expensive as in Japan. For another, American manufacturers could never fully trust their suppliers and dealers as well as the Japanese could. Relationships between suppliers and manufacturers were often as adversarial as those between management and labor.

The story is often told that in the 1930s and 1940s one could walk through the plants of the Big Three auto companies in Detroit and literally smell food emanating from the washrooms. Relations between labor and management had deteriorated so seriously that management promulgated one emasculating rule after another and in turn workers flaunted their contempt for management by running, on company time, small restaurants and barber shops on plant facilities.

The divisiveness between labor and management was another problem that Japan had learned to avoid by forging more benign, less acrimonious relations between all segments of their society. America had always been the quintessential land of *in*dependence; Japan, of *inter*dependence. Because American factories could not count on the right parts, let alone ones of high quality, being delivered to the right place at the right time, they accumulated raw materials and finished products so that they could ride out whatever ups and downs appeared in the input as well as output sides of the manufacturing process.

As incredible as it seemed at the time, the United States' huge domestic markets did finally become saturated for traditional goods. True, a continuing market always exists for refrigerators and the like. But such markets are mostly mature by now; they're not the same as they were at the introduction of a new device. More important, consumer tastes have become so finely tuned and differentiated that it's becoming all but impossible to sell one kind of anything to everybody. The explosion in styles and varieties of running shoes is just one example. Who could have foreseen that the basic black-and-white gym model of the 1950s would have subdivided into so many different types and styles?

Volatility

As if the proliferation of different varieties of products were not enough, their life cycles began to shrink virtually to zero. With the increased sophistication of mass-marketing techniques, products diffuse faster and faster through the entire population. But this raises an incredible dilemma for all businesses. How does one plan for products whose life cycles may be considerably shorter than the developing or production times involved in making and getting them to the marketplace? How does one ensure that a market will still exist for one's products by the time they are delivered to consumers? What, in short, does one do when tastes shift faster than the times involved in planning, producing, and distributing products? (The "answer" is that knowledge of the key assumptions on which one's basic strategies depend is extremely critical. One's strategies are only as good as the assumptions one makes— business strategies must anticipate radical shifts in basic assumptions.)

With the spread of the modern computer, information not only traveled freely, but so did knowledge. The undeveloped economies of the Third World, as well as the smaller economies of our allies, learned some invaluable lessons. Instead of exporting their raw materials to us and importing back final end products, they could instead buy the necessary technology to convert *their* raw materials into finished products on their own shores. They could then sell the products cheaper to the rest of the world, including themselves, than we could. Their cheaper labor costs were more than enough to offset the cost of transporting goods long distances. Other societies learned from our experience, and their learning increased exponentially. We were no longer the sole leading center of technology and management expertise. The Japanese, Koreans, Taiwanese, and Brazilians, among many, learned how to design and manage entirely new kinds of industrial enterprises for which we were no longer the models. We forgot one of the critical lessons that we started from on the path of industrial prowess. In the beginning, we copied the British much like the Japanese aped us. We then used our Yankee ingenuity to tinker, to improve on, the inventions of others much as others were now implementing inventions made by us. We were struggling to learn a painful lesson; we were no longer the center of the universe, a lesson no former empire ever finds easy.

A Giant Machine

Of all the advantages we enjoyed immediately following World War II, perhaps none was as critical to the disadvantages we now confront as one in particular: bigness. Whereas once bigness in itself connoted and led to further strength and advantage, increasingly bigness resulted in inefficiency, ineffectiveness, and even weakness. What happened was that the underlying infrastructure of the world changed so radically that in effect the basic rules of the game changed altogether.

Consider the matter this way. If the world were indeed a machine, and a simple one at that, then its understanding, design, and operation would be rather straightforward. Hadn't the Industrial Revolution taught us (better yet, convinced us) that everything in the world resembled a machine, including humankind itself, and hence was explainable in machinelike terms? Since one of the prime properties of machines is that they can be broken apart and hence ultimately built from decomposable parts, wasn't the same therefore true of humans and all their institutions? Wasn't the human body also decomposable into separate component parts or organs that had nearly autonomous existences? Further, wasn't knowledge itself subdividable like the different parts of a machine so that it could be neatly codified into separate autonomous disciplines and professions? And, finally, didn't we learn from the great German sociologist Max Weber that organizations themselves were machines so that if one part failed to work well, one merely had to pull out that defective part and replace it with a new one?

One of the most appealing properties of the world if it were a machine is that the bigger the inputs into it, the bigger the resulting outputs. And, generally speaking, this is true *if* the machine is able to maintain the same level of output efficiency no matter what the size of the material put into it.

Unfortunately, human institutions do not fit this general pattern. What we find increasingly on every front of human existence is that greater or larger inputs into human organizations do not result in greater outputs or benefits. In order to handle greater inputs, human organizations generally must become bigger, and bigger organizations are usually more inefficient than smaller ones. (This is not always necessarily true, nor does it have to be, but if and only if we are willing to expend greater energy in getting the organization's members to work together better, the larger the organization becomes. Without greater cohesiveness and understanding between members, the larger an organization gets, generally the more inefficient it becomes.) Thus, the mergers of more and more organizations into fewer bigger ones that in turn begot even bigger organizations right after World War II led to the general increased *in*efficiency of all organizations, private as well as public.[6]

When bigness became coupled with the world's increased interdependency, which in turn was aided by the modern computer revolution, the final result broke once and for all the lock that the simple conception of the world as a machine held on society's mind. It is well known that all processes have an upper limit to their efficiency. At some point bigness boomerangs on itself. Instead of leading to greater end benefits, it produces negative end effects. These negative benefits are almost impossible to foresee and go completely counter to the hypothesized benefits of bigness. Thus, instead of "bigger being better" or "more inputs into the machine leading to bigger desired outputs," more or bigger inputs into an inefficient machine generally lead to less.

The Loss of Slack

The point can be summarized in a single proposition: *The United States has lost, perhaps once and for all, whatever huge slack and buffering was built into the system as it existed immediately after World War II.* When the United States had huge unsaturated domestic markets that were hungry for anything that could be produced, the United States could thus get away with equally huge bureaucratic organizations and production lines that were sloppy or inefficient. We could tolerate friction and hostility between labor, management, government, and stockholders. Today we can't get by any longer with this kind of behavior. We're competing today with countries that make quality goods because they have forged close *alliances* between their employees, managers, governments, and shareholders. They may not be perfect alliances, but they stand in sharp contrast to our adversarial relationships. In short, the United States was shielded from the rest of the world by temporary advantages that have now passed to others.

[6]M. Harris, *Why Nothing Works, The Anthropology of Daily Life* (New York: Touchstone, 1981).

What Can Be Done?

What, if anything, can the United States do to improve, if not regain, its position? What is the role of knowledge, of *new thinking,* in this effort? What are the basic ideas on which *old thinking* rested, and why is old thinking no longer appropriate? If new thinking is required, what does it involve?

The rest of this book offers answers by presenting various philosophical systems as problem-solving or knowledge systems. It shows the strengths and weaknesses of each and the kinds of problems for which each system is best suited. We contend that, with few exceptions, professional schools largely teach their students how to solve bounded-structured problems. Such problems are generally of the type "$x + 5 = 11$, find x." The problem is "structured" in that it is phrased unambiguously in a language, in this case simple algebra, that states clearly what the problem is and gives an equally clear procedure for finding the solution, $x = 11 - 5 = 6$. Just as strongly, there is a clear sense of what constitutes a "solution." The problem is "bounded" in that there is a finite set of appropriate "solutions" to the initial problem. In this case, there exists only a single solution, $x = 6$.

Unstructured problems, on the other hand, are generally on the cutting edge of knowledge. In the early stages of research, there may well be no single, accepted way of posing or structuring a problem to the satisfaction of all experts. Further, unbounded problems such as "What is justice?" appear at times to have a literally infinite number of solutions.

Unstructured-unbounded problems are the most fascinating. Most, if not all, social problems are of this kind. We will describe a way of analyzing such problems that we call Unbounded Systems Thinking. And we will show how this new way of thinking can be applied to a wide range of real-world problems.

Plan of the Book

Chapter 2 examines the role of Agreement as one of the first and classic ways of knowing. Chapter 3 discusses Analysis or mathematical model building as one of the other important and traditional ways of knowing. Chapter 4 is one of the first systems that marks a transition between the old and new ways of thinking. It introduces the concept of Multiple Realities or the necessity of viewing all important problems from multiple viewpoints, although it does not specify a "general method" for doing so as Chapter 6. Chapter 5 describes the concept of the Dialectic or the necessity of Conflict in analyzing important problems. Chapter 6 introduces Unbounded Systems Thinking, which we contend is necessary for the problems we face. Chapters 7 and 8 apply Unbounded Systems Thinking to actual decision situations. The experience of the disaster at Bhopal, India, is an especially vivid example of the failures of outmoded thinking. A final chapter examines some of the philosophical implications of this new way of thinking.

PART I

OLD THINKING

CHAPTER
2

Agreement:
The First Way of Knowing

PROOF, n. Evidence having a shade more of plausibility than of unlikelihood. The testimony of two credible witnesses as opposed to that of only one.

RATIONAL, adj. Devoid of all delusions save those of observation, experience and reflection.

REALISM, n. The art of depicting nature as it is seen by toads. The charm suffusing a landscape painted by a mole or a story written by a measuring-worm.
Ambrose Bierce, *The Devil's Dictionary* (New York: Castle Books, 1967)

How to manage the Himalayas? That, we soon realized, was the question we had been handed. We've sought the help of a school of management. "Ah," said the head of the school. "That is not really a management problem. Management, as we define it here, is management *within* an organization." Since there is not nor is there ever likely to be a Himalayas Ltd., the sort of mess we had been handed was, in terms of management science, something very unpleasant indeed. It had to do with *inter*-organizational decision making.

The management scientists were quite right to back away from us. They tend to work in (or as consultants to) organizations, and they tend to work with those who are towards the top of those organizations. In such culturally homogeneous settings the chances are that everyone will see the problem much the same way (if only because anyone who does not soon finds himself on the outside of those organizations). Organizations maintain themselves by maintaining cultural homogeneity; they generate *shared understandings* of the problems they face. Each problem, in consequence, has but a single definition. . . .

What has impressed us most in our survey of the extensive data available on "the problem" in the Himalayas is the remarkable range of expert opinion on a number of key variables. Whether the subject of enquiry is fuelwood use, agricultural production, or even whether there is a connection between deforestation and flooding, the collection of valid and non-contradictory data is extraordinarily difficult.

The available data, in other words, do not suggest a single credible and generally consistent picture of what is happening in the Himalayas.
Michael Thompson and Michael Warburton
"Decision Making Under Contradictory Certainties: How To Save The Himalayas When You Can't Find Out What's Wrong with Them"
Journal of Applied Systems Analysis, Vol. 12, 1985, pp. 3–4, 11.

Consider the following scenarios:

> Scene 1—You enter a room. In it are four of your peers, four other U.S. Senators. All have agreed to intervene on behalf of an important member of the U.S. savings and loan industry. You have serious reservations regarding the propriety of the meeting to begin with, let alone with the particular case under discussion. Do you go along with the others, especially since you are being strongly pressured to do so?

> Scene 2—You are a member of the President's National Security Council. A U.S. Marine colonel acting presumably on the orders of the President urges you to approve a covert action which, if made public, will put you personally in jeopardy. You are pressured by other members of the National Security Council plus other advisory members of the President's staff to go along. What do you do?

Data Are Not Information

Almost without exception, all who write about the new, global information age acknowledge that we are literally drowning in an overload and overabundance of information. Never before has humankind had access to so much, so quickly, and from every part of the globe. We have more data and information on every conceivable subject, yet less understanding at the same time. Data and information do not automatically lead to greater insight; they may now travel at the speed of light, but understanding and wisdom do not.

There is also common agreement that "data," "information," and "knowledge" are not the same, even though they are often—wrongly so—used interchangeably. Their differences are often as unclear to the experts as to the layperson.

One aspect above all else is especially disturbing. It is the strong, taken-for-granted assumption that *agreement* between the monumental and voluminous databases that both government and business are constantly producing will eventually result, and further, that agreement itself is fundamentally desirable. This common but misguided philosophical assumption is the starting point for this chapter. It critiques one of the major philosophical bases on which professional education rests, the philosophy of empiricism.

The Tale of the Delphi

To show how the philosophy of empiricism enters into professional education, let us begin by considering a deceptively simple question—

one inordinately easy to state but devilishly difficult to answer: What are the respective tonnages of steel that Japan versus the United States will produce by the year 2000? In one form or another, thousands of such questions are asked daily by business and government organizations. The answers to these questions, both real and hypothetical, are the stuff of countless important decisions. They are used to develop new plants, products, marketing campaigns, and even whole new businesses.

In this chapter, we consider one of the major historical ways or traditions of approaching this or any other question. Of all the means that humans have invented to learn about and control their world, none has been more persistent or more powerful than the notion that for something to count as knowledge, it must be based on facts or observations. Furthermore, those facts or observations must be such that in principle "all competent observers" can observe the same facts.

The notions of agreement and observation are so fundamental that they constitute one of the major cornerstones of modern science:

> The objectivity of science depends wholly upon the ability of different observers to *agree* [italics ours] about their data and their processes of thought . . . In the last analysis *science is the common fund of agreement between individual interpretations of nature* [italics ours]. What science has done is to refine and extend the methods of attaining agreement. . . .
>
> All that can be claimed for science is that it focuses upon those *primary observations* [italics ours] about which human observers (most of them) can *agree* [italics ours] and that it emphasizes those methods of reasoning which, from empirical results or the successful fulfillment of predictions, most often lead to mental constructs and conceptual schemes that satisfy all the requirements of the known phenomena.[1]

The difficulty with the steel question in the preceding paragraph is, "What facts or observations can we make about that which has not yet occurred, that is, the year 2000?" If in order to know something by the approach we are describing, one has to be able to make direct observations, then one cannot know the future by this approach at all. Yet this is not the conclusion reached by those who have advocated this way of knowing.

There is a particular approach that the latter part of this century has developed that reveals both its nature and its perennial appeal. It is called the Delphi method after the famous Greek oracle.[2] While the Delphi is certainly not the only or the best approach that could be used

---

[1]Bentley Glass, "The Ethical Basis of Science," *Science,* December 3, 1965, Vol. 150, pp. 1254–1261.

[2]See Harold A. Linstone and Murray Turoff (Eds.), *The Delphi Method, Techniques and Applications.* (Reading, MA: Addison-Wesley Publishing Company, 1975).

to illustrate the particular way of knowing that is the subject of this chapter, it clearly reveals many of its most distinctive features. The Delphi also illustrates something equally important: the many attempts to formulate this particular way of knowing in scientific or quasi-scientific terms.

The Delphi procedure has many variations.[3] Essentially, the method consists of putting a question (in this case the expected tonnages of steel produced by Japan versus the United States in the year 2000) to a panel of experts. Typically, the experts are dispersed in different time and geographical zones. Each is asked to give his or her best individual estimate of the respective tonnages. The *range of responses* are compiled and transmitted to each panelist. Each expert is then asked for a second estimate that may be the same as the first or different. The process is repeated over additional rounds. It usually stops when the responses become stable, that is, show little change between rounds. In fact, the convergence of responses is common, but not necessarily inherent in the procedure. If it is done properly, the Delphi may expose divergent views and exhibit the uncertainty that exists within a large group of experts.

Three main characteristics of the Delphi are: (1) group participation, (2) the iteration of responses over various rounds, and (3) the anonymity of responses. The aim is to avoid the tendency in conventional group meetings for certain individuals (superiors, seniors, prestigious figures, opinion leaders, etc.) to sway or inhibit others, while still gaining the advantage of collective expert input and interaction. In this hypothetical case, the output will be two distributions of estimates representing the range of views of the expected tonnages of U.S. versus Japanese steel production in the year 2000.

Many variations on the basic procedure described above are possible. For example, in the second round, a Delphi might include a request for the panelists to give their reasons if their first-round estimates departed significantly from the mean of group scores. This of course sends a signal to potential mavericks that they are indeed "mavericks." This may then induce them to converge to the group mean. Thus, the Delphi may also exert pressures for individuals to conform.

In the third round, a tabulation of the various reasons would then be fed back anonymously to all panelists together with the distribution of scores from the second round. Sometimes a maverick with a disparate estimate has based it on significant reasons that are ignored by the others. Thus, this procedure allows the group as a whole to give at least

[3] *Ibid.*

some consideration to the reasoning of the mavericks before forming judgments for the next rounds.

The desire for agreement is normal in all social intercourse. It motivates committee chairpersons to persuade, cajole, even force group members to change their minds. It is in evidence even when no overt pressure exists, as in the typical Delphi. Furthermore, as in the case of committee chairpersons, the Delphi organizer may cut off "outliers" or maverick responses to facilitate, if not produce, convergence. Although such actions are frowned on at the level of free individual expression, as well as in the typical Delphi design, they do nevertheless reflect the strong philosophical bias toward agreement that is often seen as vital at the collective level.

To illustrate how such outliers may be excluded, we will focus on a particularly extreme version of the Delphi in which "outliers" are omitted from subsequent rounds. This extreme case is not typical of the Delphi procedure in general but we examine it here because it depicts one of the prime characteristics of the particular way of knowing discussed in this chapter. Other less extreme variations of the Delphi are illustrative of some of the other ways of knowing that we will describe in later chapters. (However, as we will see, there are no moderates in the "land of knowing." *All ways of knowing are embodiments of extreme principles of one kind or another. Every one of the different ways of knowing that we examine in this book is an "extreme."* In addition, each "extreme" is extremely different from each of the others.)

An important number that all students of statistics learn to compute can be used as a basis for the decision regarding which experts to exclude from successive Delphi rounds. This number is the "standard deviation" that measures essentially the "spread" of a set of numbers or estimates. For instance, if we have three experts and their respective estimates are 1,000, 2,000, 3,000 tons (thus the average equals 2,000), and another group of experts whose estimates are 1,000, 2,000, and 100,000 (the average thus equals 34,333), then the spread of the estimates of the second group is significantly greater than that of the first. The standard deviation can then be used as a criterion or cut-off point by which to keep or to exclude experts from round to round.

Figure 2.1 depicts a typical situation where the estimates of experts vary systematically around an average of 100,000 tons. Figure 2.1 portrays the famous "bell-shaped" curve. The height of the curve at any point gives the number of experts who agree with a particular estimate. In the case of Figure 2.1, the height is largest at the average. The greatest number of experts believe in this hypothetical case that U.S. steel production will be 100,000 tons in the year 2000.

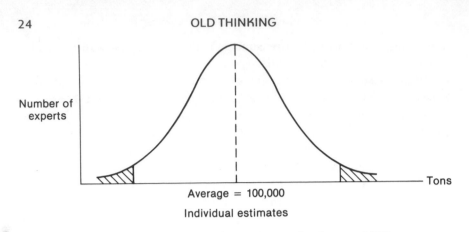

Figure 2.1 U.S. steel production in tons for the year 2000

The extreme form of the Delphi reveals that: first, the *average* of a set of numbers or estimates is used as a measure of "the truth"; second, the *standard deviation* or the spread of a set of numbers is used as a measure of the "strength" or "goodness" of the truth. Thus, if the averages from successive rounds of the Delphi are converging toward a single, stable number, or a set of numbers, and, further, if the standard deviations from successive rounds are getting progressively smaller and smaller, thus indicating that the experts are coming closer and closer to common agreement in their individual estimates, then the average is taken by this procedure as a reasonable measure of "the truth."

Does the Delphi in general, let alone the extreme version we have been exploring, produce what its proponents claim it does? To answer this question, we must examine why the Delphi came into being in the first place. That starting point is one of the most famous experiments in the history of social psychology, known as the Asch effect.[4] If Solomon Asch, after whom the effect is named, had done no other work in his long and distinguished career, his name would still be enshrined among the great psychologists.

The Asch Effect

Asch was concerned with the issue of conformity, the conditions and circumstances under which individuals would bend their judgments to those of others. If humans are anything, they are deeply social animals.

[4]S. E. Asch, "Effects of Group Pressure upon the Modification and Distortion of Judgments," in Harold Guetzkow (Ed.), *Groups, Leadership, and Men* (New York: Russell & Russell, 1963), pp. 177–190.

They are not isolated atoms or machines that form impressions about the world uninfluenced by others. They affect one another continuously in countless subtle and not so subtle ways. Asch explored these influences in one of the most simple and elegant experiments imaginable.

Surely, if there is any one thing that is supposedly *not* open to the influence of others, it is the clear and decisive testimony of our senses. Asch tested this basic premise and found it wanting.

Asch showed the subjects in his experiments a screen containing two different lines, A and B, that were substantially different lengths, say, in the ratio of 1.5 to 1.[5] Thus, line A, for instance, might be 1.5 times larger than line B. Subjects were then asked whether the two lines A and B were the same in length or different. Subjects who were tested individually in isolation were able to state decisively that the two lengths were clearly not the same. Then the heart of the real experiment began.

Asch took aside six people and made them confederates. They were instructed to say that the two lines were equal in length even though they were demonstrably different. Next, an unwitting subject was introduced into the previously formed group of confederates. Before giving his or her judgment on the length of the two lines, the naive or uninformed subject had to listen to the judgments of the first six subjects.

The first subject was asked by the experimenter, "Bill, what do you think: are the lines A and B the same in length or are they different?" Without hesitating, Bill says that they are the same. At this point the naive subject, call him Tom, says to himself, "Is this guy nuts or what; is there something wrong with his eyes?" Next, the experimenter asks the same of the second subject, Susan. Susan says the same as Bill, "The two lines are the same." Tom is really getting disturbed by this point; he says to himself, "What's wrong with both these people; can't they see?" The process repeats itself with the next four subjects all saying that the two lines are the same length. In fact, by the time the third or fourth confederate in the experiment has agreed with the prior group, the Toms of the world show visible nervousness and anxiety. They squirm noticeably in their seats.

What do these Toms do when it is their turn to respond, after the first six subjects have given their judgments? Surprisingly, some 58% of the subjects say that the two lines are the same! Perhaps the most shocking and significant finding of all was that, although very few in number, some subjects actually came to believe that the two lines really were equal.

---

[5]Asch's actual experiments were a bit more complex. However, this discussion captures the essential features.

This experiment has been repeated countless times with innumerable variations and essentially the same obtained results, although there are important differences depending on the type of question asked. Thus, if the question is one that calls for an opinion, for instance, "Was Ronald Reagan a good President?," then naive subjects are less likely to bend to group opinion.

If humans can be pressured by their peers in so simple a social setting to disregard "the clear and objective testimony of their senses," and in cases where supposedly the testimony of one's senses is indisputable, then what must they do when pressured by their boss or other superiors to go along with actions they don't approve of? What would one do in the two scenarios mentioned at the start of this chapter? While we are not excusing this behavior, it is entirely understandable how humans can be pressured to go along in such situations. Because resistance is indeed so difficult is precisely why it is often the mark of high character. If education has any power at all, then ideally it can show us why the need to conform is so strong in every one of us, and what we can do to counteract it.

The significance of the Asch experiment is that it demonstrates clearly that people are not mere objective recording machines or cameras. This is indeed the whole point of the experiment. Two lines *objectively* different in length are presented for supposedly our eyes and our eyes alone to judge. Yet when this simple judging task is done in a social context, a substantial proportion of the population are content to bend their judgments. Further, a tiny percentage actually come to believe that the group is right. Believing can thus truly be seeing.

This is not the way the world is supposed to work, at least not according to the philosophy that goes by the name of empiricism. Although many different brands of empiricism exist, what they all share in common is the notion, however expressed, that all knowledge not only arises out of the impact of objective reality on our various senses (sights, sounds, colors, touch, tastes, smells), but that everything we pretend to know must be reduced to these and these alone if something is to count as knowledge. The assumption is that humans are essentially passive receptors of hard, objective impressions or facts from the outside world.

If the Asch experiment teaches us anything, it is that human beings do not fit this description. Humans are not mechanical, blank tablets on whom impressions are showered from the outside world and recorded passively. Even if they were, they are not isolated, independent ones ready for imprinting. Our tablets, so to speak, are highly dependent on those of others. Of course an empiricist might well contend, "Oh yes, what and how you interpret what you have *already taken in* (seen, felt, tasted, smelled, etc.) may be influenced by others, but certainly not the basic raw facts or stuff of experience itself." Although the Asch experiment does not demonstrate it, other experiments reveal just as strongly that what humans take in through their eyes and other senses in the first place is not only influenced by the social context in which they are embedded but their emotional state as well.[6] Thus, the particular experimental facts, on which empiricists are always so insistent that we should base our knowledge, do not support their own case in this matter.

One aim of the Delphi is clearly to get around the Asch effect. If the judgments individuals give about important issues can be so easily swayed by group pressure, then why not remove that pressure through geographical isolation and anonymity and see if groups can thereby make better judgments. However, the misuse of the Delphi by lopping off, or ignoring, experts whose judgments are too far removed from the average sends a blatant signal to others to conform or be ignored in later rounds. But even if no such arbitrary censorship were practiced, strong tendencies or pressures toward convergence nonetheless exist. Thus, less extreme versions of the Delphi still reflect the strong philosophical bias that is inherent in the way of knowing we are describing in this chapter.

A recent demonstration of this bias has been carried out by Rowe, Bolger, and Wright.[7] When "correct" majority feedback was given to panelists, individual predictions became more accurate in subsequent rounds. However, when "incorrect" or false majority feedback was presented to the individuals, their predictions became less accurate in subsequent rounds. In other words, individuals were drawn to the group norm regardless of the external correctness of that norm. The effect of groups on individuals reflects a relationship well recognized by media experts and demagogues.

Delphi's "Truth"

What is the meaning of "Truth" as it emerges from such a procedure? If we value agreement as a basis for action on important social matters—

---

[6]For instance, suppose we use a special optical device to flash two different pictures, a baseball player and a bull fighter, at the same time. The catch is that the baseball player is flashed to one eye and the bull fighter to the other. Also, if we decrease the time during which the two pictures are flashed, then the brain has to make a decision regarding which picture to see or to record. It should come as no surprise that Spaniards typically report seeing the bull fighter whereas Americans see the baseball player. This and other experiments thus demonstrate the strong influence of culture and social context on perception.

[7]G. Rowe, F. Bolger, and G. Wright, "The Delphi Method: An Investigation of Richness of Feedback and Change in Majority/Minority Opinions," *Technological Forecasting & Social Change*, Vol. 39, No. 3, May 1991, pp. 235–252.

and most societies do—then *one cannot act on agreement alone without knowing a great deal about the manner and the conditions under which it was achieved.* Otherwise, such agreement is not only superficial, but can be misleading, downright dangerous, and even evil, especially if it was produced under conditions of coercion.

Because our discussion has been philosophical, the reader is strongly cautioned to understand that practical consequences do indeed hinge on it. We are a nation of approximately 250 million people. We are a blend and a melting pot of all the world's cultures as no other nation is or has been. We possess neither the homogeneity nor the uniformity of the Japanese. Agreement or consensus on matters of importance—save those of the most extreme crisis situations—is thereby much more difficult for us to achieve. If anything, we always seem to be literally falling or flying apart at the seams; the center is always in danger of not holding. That the Japanese and others are able to reach consensus—that they can even achieve it at all—is an absolute marvel. We regard them with jealousy and envy, if not hostility, and ask ourselves, "Where have we gone wrong; why are we no longer able to compete economically with them and with others?" If agreement and consensus are the base on which industrial might and power now rest, then our traditional rivalries between labor and management, government and private enterprise, seem destined to make us into a permanent second-rate or inferior economic power. Hence, the ability to agree on important issues is not just an issue for esoteric philosophical debate. It bears directly and heavily on economic success.

On the other hand, disagreement can also be a strength if it fosters creative individualism. The question, of course, is, "How do we combine the best of both individualism and consensus?"

Other vital issues such as "How should we educate our children?" are also involved. What do they need to know to prosper in the new information or systems age? Obviously, all citizens must have a common core or body of facts so that they can function in a complex world, but precisely "whose facts" do they now need? The "facts" of the book, *Cultural Literacy?*[8] Of Western culture in general? Of Asian, African cultures? The "facts" of feminist scholars as opposed to those of traditional white male, Protestant scholars?

Probing further still, if widespread agreement or consensus is necessary before we act on important matters of social policy, what are we to do in important and certainly not rare cases, such as the abortion controversy where the two sides are so polarized that the very possibility of

---

[8]E. D. Hirsch, Jr., *Cultural Literacy: What Every American Needs to Know* (Boston: Houghton Mifflin, 1987).

consensus between them is impossible? *If tight agreement or consensus is required before we can act on important social issues, then we seem doomed to perpetual inaction or widespread dissatisfaction no matter which course is taken, a condition in which we have found ourselves too readily in recent years.*

Or, has our brief review of those philosophical traditions that are founded on the necessity of agreement between different observers, observations, judges, judgments, facts, and the like pointed to another conclusion? While agreement or consensus is *desirable* in many, if not most, important matters of social policy, it may not always be necessary. It certainly is not sufficient. Because all the experts agree we are winning the war on drugs, are we truly? Because all are agreed we should go to war with Iraq, should we?

While agreement or consensus may be desirable as necessary conditions before we act, it becomes all the more important that we know whether agreement or consensus was produced artificially by the method used to reach it. Like congressional votes, the Delphi can be used to produce the illusion of widespread consensus. The more general point is that all methods of producing consensus may be artificial, although not necessarily all to the same degree.

We have dwelled at length on the role of agreement and facts because they underline so much of our culture's ideas on how proper decisions should be reached. In the next section, we want to describe the particular method we have been discussing in more precise terms so that we can not only see some of its major characteristics more clearly, but can compare it more systematically with other methods.

Inquiry Systems

The Delphi is a simple but a very powerful and clear example of a particular kind of Knowledge or Inquiry System. This is in fact precisely why we have dwelt on the Delphi; our interest is not in it per se, but rather in the general philosophical attitudes that underlie it. The Delphi helps us to see these attitudes much more clearly than almost any other example we could give. In this sense, if the so-called knowledge or information revolution we are living through is a demonstration of anything, then it is that we are changing profoundly, and need to change even more, the Inquiry Systems that we have traditionally given our allegiance to in the past.

An Inquiry System, or IS for short, is a system of interrelated components for producing knowledge on a problem or issue of importance.[9] Since the IS's

---

[9]See C. West Churchman, *The Design of Inquiring Systems* (New York: Basic Books, 1971).

we describe in the following chapters differ radically in almost every one of their critical components, we want to understand why by comparing them systematically. Further, since the term "system" is by now so overworked that it is almost meaningless, we want to show that it still has important and valid applications.

First, every IS has or accepts distinctive inputs from the "outside world" (see Figure 2.2). Those inputs are not only the basic entities that come into a system in the first place, but are much more fundamental. Inputs into an IS are the entities that a particular IS recognizes as the "basic, legitimate building blocks or starting points for knowledge." *The inputs that a particular IS recognizes as legitimate are not necessarily recognizable by other IS's.* The valid starting points or legitimate building blocks of knowledge of one IS are the *in*valid and the *il*legitimate starting points of another. In the case of the Delphi, or more generally still empiricist IS's, the basic building blocks or inputs are raw facts, observations, or the various judgments of experts.

Second, different IS's employ different kinds of operators. The operator in an IS is the mechanism that operates or works on the basic inputs to transform them into the final output of the system, or knowledge. In the Delphi, the simplest operator is the averaging function, that is, the arithmetical process of adding up the separate judgments of individual experts and then dividing them by the total number of experts (e.g., in order to get the average for U.S. steel production in the year 2000).

Third, the output of an IS is what the system regards as a valid knowledge for action on an issue of importance. Thus, in the case of the Delphi procedure described in this chapter, the simplest output is the average itself—a single number that expresses the best single estimate of U.S. (or Japanese) steel production in the year 2000.[10] Notice the difference between the operator and the output. The operator is the *process* of *producing* the average—that is, the arithmetical process of adding up the individual estimates and then dividing by the total number of experts—while the output is the *average* itself—that is, a single number.

Fourth, the most critical component is known as the *guarantor*. The guarantor of an IS is the component guaranteeing the operation of the entire IS itself. Thus, the guarantor specifies (argues) why in order to obtain valid knowledge (1) one should start with a particular kind of input, (2) use a particular operator to transform it into, (3) a particular

---

[10]It is important to know that the average is not always the typical output of every Delphi. Indeed, the most typical output is the range of scores itself. Only in the more extreme example of the Delphi that we have been discussing does one find the average as "the output."

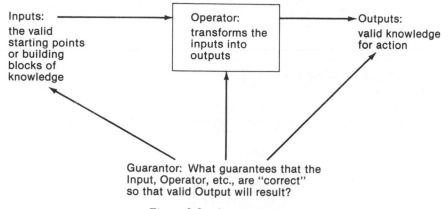

Figure 2.2. An inquiry system

form of output that is regarded as knowledge. In the case of the Delphi, the guarantor is composed primarily of two major parts: (1) the process of computing the standard deviation or the spread in a range of expert judgments, and (2) the specification of how small the standard deviation should be so that the output can be regarded as "truth."[11] In the extreme version of the Delphi described here, which is not typical of all Delphis, the second aspect of the guarantor also enters into the operator in that experts who deviate too far from the average will be used to remove those experts or mavericks from the initial pool of "valid experts." A critical part of the guarantor's job also involves selecting the initial pool of "experts" themselves. The guarantor thus contains an implied definition of who or what an expert is to begin with. For example, it is not clear that the experts should come solely from upper management. It would be interesting, to put it mildly, to see what the estimates of steel production are by groups of workers in both the United States and Japan. If one believes that the workers themselves are closest to the actual functioning of plants or factories, then they might have a deeper insight about what is truly going on. They might indeed be the relevant "experts" in this particular case.

The guarantor is the most critical aspect of an IS because it literally influences everything it does. Thus, as we will see, the differences in the often heated battles between different IS's, or philosophic systems, over what is truth, what is a proper way to decide important issues, is usually over the guarantor, although it is rarely put in this manner. Philosophic wars would be better termed "guarantor wars."

---

[11]In the most general terms, the guarantor is the belief that truth can be obtained from any Delphi procedure by using a series of experts and combining their responses.

Other aspects of an IS, more often than not, are implied rather than stated explicitly. These include:

- A preferred definition of objectivity.
- A preferred definition of education.
- A preferred measure of performance for evaluating the output of a system.
- The mood in which the system expresses itself, that is, its key findings and propositions, arguments.
- A notion of systems separability.
- The number of views of a problem that it presents formally.
- The type of problems for which the system is best suited.

If it were not so important, it would be amusing to observe how the term "objectivity" is used in our culture to squelch opponents and put down opposing arguments. The implication is that there is one and only one definition of objectivity recognized generally so that the person who violates it is thereby soundly dismissed. The "*fact* of the matter" is that there is not a single universal definition of objectivity shared or embraced by all IS's equally. One of the reasons why the study of IS's is so important is that it reveals clearly and systematically the *range* of ways in which critical terms are used, and hence differ, as we go from IS to IS.

In empiricist systems in general, something is objective—an action, conclusion, decision, hypothesis, idea—*if and only if* it can be clearly and firmly grounded in facts or observations of some kind. An action, conclusion, etc., that cannot be traced back to a firm foundation of facts or observations is not only suspect but in danger of being dismissed by the derogative label "subjective."

The preferred mode of education that IS sanctions follows directly and strongly from its preferred notion of objectivity. Thus, in the particular system at hand, education is to be slanted strongly toward the learning of facts—generally what has been validated through repeated observation or passed down from generation to generation. Theories may come and go, but presumably hard facts or observations are "forever."

The mood in which empiricist IS's speak or put their recommendations is the indicative mood, that is, something "is" or "is not." The mood and measure of performance are also strongly related. If at all possible, the measure of performance is expressed in the form of a single number or numbers that represent how well the system has performed in the past or is performing currently. In the particular case of the extreme version of the Delphi, the measure of performance is the standard deviation since it measures the tightness or the strength of agreement between the set of different experts.

Next to the guarantor, the concept of systems separability is significant. Every IS has some things that are under its direct command or purview and those that it takes for granted or presumes and are thereby outside the system. In the case of the Delphi, the system takes for granted that an expert is someone who can express a judgment in the form of numbers. How the experts get their numbers in the first place is not of direct concern. The numbers merely come in from the "outside world of the experts," much like sense data just arrives on the scene for empiricists. Thus, what assumptions, biases, concepts, models, etc., that the experts may have used to produce estimates in the form of numbers is neither of concern nor of direct interest to this IS. The process of producing the numbers is *separable from* the process of treating or operating on the numbers to extract "truth" from them. For those not caught up in the spirit of this particular system, this makes absolutely no sense, as do many of the system's other features. But in general this is true of every IS. What is reasonable and makes sense from the perspective of one IS makes absolutely no sense from that of another.

(C. West Churchman gives a simple example of systems separability that shows how critical it is.[12] Churchman was summoned for jury duty. Prospective jurors were individually asked in turn whether they could separate their decision as to the guilt or innocence of the accused party in the particular case at hand from the rest of the criminal justice system. One by one, the jurors said they could. Churchman could not, for the question that haunted him was, "What if the rest of the system was inequitable? How could I judge a person guilty or not independently of the rest of the system, that is, whether it was functioning properly, or whether it was designed properly in the first place? Don't I have to make strong assumptions about the fairness of the rest of the system in order to perform my function in this particular case at hand?" Whether one agrees with Churchman's decision is not the important issue at hand. In dealing with any case, one has to make a critical decision as to where and how to draw the boundary between a certain part of the critical decision and other parts of the relevant surrounding system.

For instance, in the U.N.–Iraq war, a central issue was that of "linkage." In the language of this book, linkage is a synonym for separability. In the view of the Iraqis and the Palestinians, the issue of the Kuwaiti question is not separable from the issue of the fate of the whole Middle East. This is not to say that we necessarily agree with Iraq, but rather that the issue of systems separability constantly raises its head. Indeed, it is almost by definition a central issue in most important social decisions.)

---

[12]C. West Churchman, *Thought and Wisdom* (Seaside, CA: Intersystems Publications, 1982).

In the extreme Delphi, typically one view of a problem is presented and an attempt is made to solicit one "best answer." For instance, the initial question (What is the best estimate of Japanese versus American steel production in tons for the year 2000?) is formulated or posed in one way, and one answer is typically sought. Thus, the extreme form of the Delphi typically does not present a range or a variety of ways of asking a question. Neither does it usually modify the question over successive rounds.

The strong presumption of such systems is that only well-structured questions are meaningful—those that can be stated and answered precisely. Thus, questions and concerns of aesthetics, ethics, and values—to mention only a few—are usually dismissed by such systems since they can neither be posed from one and only one perspective and certainly do not have a single "answer" that can be put into the form of a single number. (From the standpoint of the other IS's that we will explore later, the mere thought of putting questions of aesthetics, ethics, and values in the form of numbers is almost too laughable to warrant serious analysis.)

Conclusions

Several conclusions emerge from our discussion of the first way of knowing.

1. While we have used the terms "agreement" and "consensus" interchangeably—dictionaries often confuse them as well—agreement and consensus are not always the same. For instance, the single numerical estimates of steel production of two different experts could well agree or match exactly, yet the two experts may not necessarily share consensus. The underlying reasons for their numbers may differ sharply. Whether two or more experts agree or not depends on (a) the way the initial question is asked, (b) in what form or forms the answer is stated, (c) how deeply we probe the experts for the underlying reasons behind their answers, (d) whether we allow the question, process, and answer to vary over rounds, and (e) whether and how we involve the experts in the design of the process (e.g., the Delphi itself) in which they are participants.

2. Philosophers have generally been extremely insensitive to the critical and significant role that social and psychological factors play in the perception of facts, observations, sense data, and so on. If humans were perfect machines, then the functioning of their sensory organs might be independent of the influence of others, their emotional state,

and other factors. It might also be a legitimate strategy to study how humans perceive and order their world in isolation from others. However, humans are supremely social animals. Therefore, every decision procedure designed to correct for one set of social influences or biases only reintroduces them via the back door in some other fashion. In the language of systems, social influences are never fully separable from all the other parts of an IS. Further, it is not clear that they should be. Rather than the complete elimination of social influences or biases being our goal—and an impossible one at that—our goal instead should be to understand how such influences and biases operate.[13] Unfortunately, it is not possible to study that which one thinks one has eliminated or is not important to study.

3. The designer or the "wizard" behind the Delphi can never fully be removed or made separable from a Delphi itself. Who after all has selected the initial question? Why is it of interest to him or her? Who is paying for the results? How will they be used? None of these questions can ever be fully independent of, or separable from, any serious inquiry. The academician's traditional disclaimer that he or she is only interested in a narrow set of prescribed concerns is no longer acceptable—if it ever truly was—in a world that is smaller and more interconnected. The things considered "irrelevant" or "not my concern" may be the most important things we need to know about an IS.

4. We have dwelt at length in this chapter on a very special type, or extreme form, of the Delphi because it is a prime, concrete example of an empiricist system and, hence, exhibits many essential features of empiricist systems in general. More generally, the Delphi is an excellent example of an Inductive-Consensual IS; "Inductive" because the system attempts to infer a general conclusion (e.g., steel production) from a limited set of observations. It is important to discuss Inductive-Consensual systems for, along with the IS discussed in the next chapter, they form the overwhelming core of what has traditionally been the foundation of education in Western societies—the collection and analysis of facts and observations. In recent years, the Japanese and the Koreans may have taken this even more seriously than the United States. Our point is that while it is still important for students to know and use Inductive-Consensual IS's, it will become even more important for them to learn the IS's we discuss in succeeding chapters.

5. Inductive-Consensual IS's are appropriate only for a very, very limited class of problems, that is, bounded, well-structured problems

---

[13]Ian I. Mitroff, *The Subjective Side of Science: A Philosophical Inquiry into the Psychology of the Apollo Moon Scientists* (New York: Elsevier, 1974).

(see Chapter 1) for which single numbers can serve as answers. However, the number of problems or questions that fit this mode of reasoning have come dangerously close to zero since very few questions today are amenable to this treatment. Nonetheless, the conclusion is not that we should thereby give up all reliance on facts or observations. Because the proper determination and use of facts may not be as simple as the Delphi or Inductive-Consensual IS's make it appear does not mean that there are not other more sophisticated systems using facts and numbers in more palatable ways.

6. Inductive-Consensual IS's are used widely in everyday life. For instance, the scoring procedures in ice skating and olympic diving contests are prime examples of simple Inductive-Consensual IS's. In both cases, the high and low scores are thrown out to compensate for bias (e.g., a judge from the same country as the contestant gives too high a score and a judge from another, competitive nation gives one too low). The remaining scores are either summed or averaged to produce the individual's total score for the particular event or round.

In general, the use of judges is warranted in such cases since evaluating human performance is not something that can be entrusted entirely to machines or mathematical formulae. In the case of ice skating and diving, no machine or formula can ascertain "artistic interpretation." Such aspects must instead be judged by all-too-fallible human beings. Thus, there *are* many instances in which Inductive-Consensual IS's are appropriate. Indeed, none of the IS's we examine in this book would have come into existence or survived for long if they did not have something—in many cases a great deal—going for them. The point is, however, that a procedure that is valid for one set of very special circumstances is not valid for all.

7. There is a strong sense in which the use of Inductive-Consensual IS's is undemocratic. The purpose of democracy is not to reduce important issues to polling or popularity contests nor is it to secure faster and more accurate polls of public opinion. The fundamental purpose is to have open-ended discussion of key issues, or even better, intense arguments and debates by those on opposing sides. Democracy becomes perverted when it is not only confused with consensus but reduced to numbers.

8. The reader should not be misled into thinking that the IS discussed in this chapter is the only one that makes use of agreement. Every IS has a different definition and concept of agreement and hence uses it in very different ways. As we will see, *every IS makes use of every other IS and the concepts that are embedded in them.* This does not mean, however, that every IS is consciously aware of its dependencies on every other IS.

More often, its relationship to other IS's is unconscious and therefore remains unacknowledged. Further, although every IS uses some concept of agreement, no other IS is as centrally dependent on it as the one discussed in this chapter.

9. In the end, the most general conclusion is "seek agreement or consensus but do not trust them fully." Agreement and consensus *are* important in reaching conclusions and in achieving the necessary support to carry out complex, important policies. However, as with all things human, they cannot be followed blindly. Nor are they the ultimate consideration for deciding all important questions.

> The hyper-speed of change today means that given "facts" become obsolete faster—knowledge built on them becomes less durable. To overcome this "transience factor," new technological and organizational tools are currently being designed to accelerate scientific research and development. Others are intended to speed up the learning process. The metabolism of knowledge is moving faster.[14]

> *Americans believe in facts, but not in facticity* [i.e., they do not believe or understand the general concept of or philosophy that stands behind all "facts"] [italics ours]. They do not know that facts are factitious, as their name suggests. It is this belief in facts, and the total credibility of what is done or seen, and this pragmatic evidence of things and an accompanying contempt for what may be called appearances or the play of appearances—a face does not deceive, behavior does not deceive, scientific process does not deceive, nothing deceives, nothing is ambivalent (and at bottom this is true: nothing deceives, there are no lies, *there is only simulation* [italics in original], which is precisely the facticity of facts)—that the Americans are a true utopian society, and their religion of the *fait accompli* and the naivete of their deductions, and their ignorance of the evil genius of things. You have to be utopian to think that in a human order, of whatever nature, things can be as plain and as straightforward as that. . . .[15]

[14]Alvin Toffler, *Powershift* (New York: Bantam, 1990), p. 427.
[15]Jean Baudrillard, *America.* Translated by Chris Turner (London: Verso Press, 1988), p. 85.

CHAPTER 3

The World as a Formula:
The Second Way of Knowing

LOGIC, n. The art of thinking and reasoning in strict accordance with the limitations and incapacities of the human misunderstanding.

Ambrose Bierce, *The Devil's Dictionary* (New York: Castle Books, 1967)

. . . While Gore in the Senate was questioning the models relied on by the go-slow camp, Sununu in the White House was challenging the reliability of the models that provided ammunition for the other side. Wrote *Insight* magazine: "He is on top of the scientific literature and thinks the computer models predicting significant [global] warming are too primitive to form a reliable base for action."

Today, whether dealing with the economy, health costs, strategic arms, budget deficits, toxic waste, or tax policy behind almost every major political issue we find teams of modelers and counter-modelers supplying the raw materials for this kind of political controversy.

A systematic model can help us visualize complex phenomena. It consists of a list of variables, each of which is assigned a weight based on its presumed significance. Computers make it possible to build models with much larger numbers of variables than the unaided intellect alone. They also help us to study what happens when the variables are given different weights or interrelated in alternative ways.

But no matter how "hard" the final output may appear, all models are ultimately, and inescapably, based on "soft" assumptions. Moreover, decisions about how much importance to assign to any given variable, or its weighting, are frequently "soft," intuitive or arbitrary.

Alvin Toffler, *Powershift* (New York: Bantam, 1990), pp. 291–292

Economic journals are filled with mathematical formulas leading the reader from sets of more or less plausible but entirely arbitrary assumptions to precisely stated but irrelevant theoretical conclusions.

Wassily Leontief, 1973 Nobel Prize Winner for Economics
Time, August 27, 1984

We have almost reached the point where, if you can't quantify it, you can't say it.

Lester Thurow, M.I.T. *Time*, August 27, 1984

In the last chapter, we examined that perennial mode of knowing that insists on grounding all knowledge in the agreement between experts, facts, or observations. We concluded that while agreement, experts, facts, and observations are important, the view of them presented by Inductive-Consensual IS's is seriously flawed. In challenging the Inductive-Consensual IS, we are not necessarily ruling out the ultimate importance and use of agreement, experts, facts, and observations. We are only questioning the account of their functioning, use, and importance as portrayed by Inductive-Consensual IS's.

It has always been one of the key tenets of empiricism, the philosophy that underlies Inductive-Consensual IS's, that certain sights, sounds, tastes, smells, sensations, or touches exist that are so forceful, so strong, so clear that they can neither be doubted nor confused. No one experiencing them could mistake them; their forcefulness commands every human to experience them without error and thus use them as the valid starting point for all knowledge. The difficulty with this notion is that it is flawed at its very core. There are no simple or basic sights, sounds, tastes, smells, sensations, or touches that humans can experience without error or social influence.[1]

Consider the experience of pain. If ever there were a candidate for a forceful experience or sensation, then surely it is pain. The experience of pain is so great—so strong—that the individual feeling it cannot supposedly be "mistaken" by its presence. And yet, sadly enough, the "data or facts of scientific experience" speak otherwise.[2]

Like every other sensation that humans are capable of, the experience of pain has been found to be highly dependent on social psychological variables. For instance, football players who have broken a limb during the heat of play report only being aware of the pain *after* the play or the game is over. During the game, their minds were focused or occupied elsewhere. What one experiences is not merely or solely a property of the thing being experienced itself, but rather the result of a

[1]This does not mean that empiricism is not the prevailing philosophy in American society. As Will McWhinney notes, "sensory reality is based [on] the belief . . . that which we touch, smell, see . . . is what is real. It presumes the existence of a *prima materia*, substances detectable directly or indirectly, by our senses that are related to each other and to humans—as part of the natural environment—'in certain ways.' This is the reality of the practical person, of science and commerce. It is the espoused belief system of the educated American, certainly of natural scientists and oddly, also of social scientists and psychologists. It is so pervasive that the majority of Americans have grown up to assert its *reality*, no matter how uncomfortable or ill-suited it may be to one's personal and ethical views." Will McWhinney, *Paths of Change, Strategic Choices for Organizations and Society.* (Los Angeles: Sage Publications, 1992), p. 39.

[2]Ian I. Mitroff, "Solipsism: An Essay in Psychological Philosophy," *Philosophy of Science*, Vol. 38, No. 3 (September, 1971), pp. 376–394.

complex interplay between the mind and sensations of what is being experienced.

It has also been found that how one experiences pain in the present is heavily dependent on the kind of family in which one was reared. In general, people who grow up in smaller families experience and react to pain differently from those reared in large families. Birth order also affects how one experiences pain. In both cases, one's experiences are highly dependent on how one learned to cope or complain at an early age. Large families provide more opportunities to complain or share one's pain than smaller ones.[3]

If the human organism were a simple machine, then the experience of any sensation might indeed be nothing more than a direct physical response uninfluenced by the social context or one's personal history. But since humans are not machines, they do not react to or experience sensations in the ways philosophers have traditionally hypothesized.

Like all of us, philosophers are still the prisoners of the Machine Age, the world of the Industrial Revolution. We forget how profound a revolution that was in uprooting humankind from an agrarian world and replacing it with a mechanical one. It involved a profound and radical change in human thinking. Where once the world was perceived in living, organic terms, it now was viewed and characterized in machine-like terms. Since by definition a machine is something capable of being broken down into its independent, constituent parts or components, so in principle was everything in nature, including humans. Further, since the parts of a machine are not dependent on purposes or the social context in which they operate, the same was thought to be true of the constituents of the human body, if not of humans altogether. Nowadays, we know that none of this is true. There is no part of the human body or condition incapable of being influenced significantly by one's overall mental state and social context. Yet the acceptance of this was long and hard in coming—and still is not fully recognized—because of the extreme hold that the view of the World as a Machine exercised on the imagination of humankind. Alvin Toffler notes:

> Most American managers still think of the organization as a "machine" whose parts can be tightened or loosened, "tuned up," or lubricated. This is the bureaucratic metaphor. . .
>
> Another key factor of shifting power on the job has to do with the concept of interchangeability. One of the most important innovations of the Industrial Revolution was based on the idea of interchangeable parts . . . workers, too, came to be regarded as interchangeable.[4]

[3]*Ibid.*
[4]Alvin Toffler, *Powershift*, (New York: Bantam Publishers, 1990), pp. 185, 212.

The World as a Formula

In this chapter, we want to examine another great perennial philosophy—rationalism or the Analytic-Deductive IS. More than either the Inductive-Consensual or Analytic-Deductive IS's wish to admit, both were shaped by a common underlying image: the World as a Machine. Both believe that knowledge starts from simple inputs or building blocks, although they differ sharply with regard to what these simple inputs are. Both also share the view that truth is singular, that there is a single, clear truth that emerges as the output of inquiry, although they differ sharply about the nature of this output. Most fundamentally, they differ radically regarding the nature of the guarantor. To see this, let us consider another seemingly simple, if not almost trivial, example.

Analytic-Deductive IS's: Decision at Zenith Life

There are so many examples of Analytic-Deductive IS's that history provides us with an abundance of choices. Perhaps *the* classic example is Euclid's geometry. Here one starts with the simplest of all inputs—fundamental ideas about space itself—that are so basic that they are "true" literally by definition. Thus, in Euclid one starts with basic postulates and axioms, for example, a "point" is something that has no width, breadth, or thickness; a line is merely a point extended in space; or, two triangles are congruent—identical—if their corresponding lengths are the same.

Since the outputs of Euclidian geometry—theorems—have proved so successful in a wide variety of scientific and practical concerns, Euclidian geometry was hailed for thousands of years as the premier standard or method for approaching all problems. One started from clear, simple concepts and from these one deduced higher order, more complex propositions (theorems) whose truth was guaranteed by the application of strict logical reasoning. Since this method was so successful in science and engineering, it was advocated in the social realm as well.

To illustrate the use and meaning of the Analytic-Deductive IS in a social realm, we'll apply it to a situation that on the surface at least is as "simple" as the question that occupied us in the last chapter. There is a somewhat dated yet classic case in the *Harvard Business Review* that provides a perfect depiction of the Analytic-Deductive IS.[5] Four men are running for the presidency of a fictitious life insurance company,

---

[5]Abram T. Collier, "Decision at Zenith Life," *Harvard Business Review*, January-February 1962, Vol. 40, No. 1, pp. 139–157.

Zenith Life. Background information on their strengths and weaknesses, families, career history, skills, and so on, is given for all four,
although we do not receive the same information for each of them.
Thus, we know more about one candidate in one category than we do
about another. Also, the history and current nature of Zenith Life itself,
its prospects and problems, its opportunities as well as threats, are
described. The central question of the case is, "Which of the four
candidates is best qualified to head Zenith Life, given both its past
history and its current condition?"

In all the years that we have given this seemingly "simple case" to
scores of students and executives, the typical response has remained
remarkably the same. Almost every student and executive—whether
they worked individually on the case or in small groups—built *a single,
simple model* that selects *one and only one* of the candidates as best for
Zenith Life. The models are virtually the embodiment of Analytic-
Deductive reasoning whether the students and executives were aware of
this or not; in most cases, they were not.

The models essentially work as follows. A set of attributes that are
characteristic of leadership is determined or specified: for instance, how
charismatic each of the candidates is; their capacity to inspire others; the
ability to formulate a vision of what Zenith Life needs to be in the
coming decade; to present one's ideas in a direct and persuasive manner
so that others will want to join on; a clear sense of ethics and the ability
to make decisions that are ethical and moral; their past job performance—job history, personality, and so on. Other variables such as
"family support" were also included. Each candidate is then scaled on
each attribute to the degree that the individual either embodies or possesses it. Typically, a score of "1" represents the absence of a particular
attribute or poor performance on it, whereas "10" indicates the complete possession of an attribute or high performance. On more sophisticated models, the attributes are weighted differently so that, for example, the category "ethics" might be rated three times more important
than one's score in the area of "past job performance." The "best
candidate" to run Zenith Life is then selected on the basis of who has
the highest score on all the attributes and their weightings.

Almost never does an individual or a group build more than one
model in order to demonstrate explicitly that, depending on the initial
assumptions one makes, not only can one specify very different leadership attributes—and hence build very different models—but, as a
result, one can select very different candidates as "best." Even rarer is
the individual or group—although this has occurred—who turns the
whole case on its head by working backwards with the presumption that

each of the candidates is "best," but for a very different kind of company. That is, suppose one starts by assuming that each candidate is "best" and then asks the critical question, "What are the characteristics of the different kinds of *companies* for which each is 'best'?" This approach thus creatively reverses the whole decision as one of specifying a *new company* to carry Zenith Life ahead in the coming decades.

In essence, nearly everyone who reads the case and analyzes it assumes, almost without question, that it is a bounded, well-structured problem. Most people believe that the attributes or characteristics of leadership are obvious or self evident; much like a machine, the phenomenon of leadership can be decomposed or broken down into its constituent parts. In addition, not only do they assume that any individual's leadership abilities can be scaled in terms of each of the separate components, but, further, that the weighted sum of scores on each component makes sense and is virtually the same as the whole phenomenon itself. To put it mildly, this is quite a body of assumptions.

While there is often much argument and heated debate between various individuals and groups over who has *the single best* or *the right* model, very few individuals or groups doubt that "out there, somewhere, *the* definitive book, expert, or mathematical model on leadership exists." In essence, the fundamental assumption is that critical human problems can be reduced to a formula, a cookbook mechanical procedure. The trick is just to find the right model and apply it correctly.

It's all that simple—or is it really? Of course not. Indeed, it is often far easier to convince people that there are no simple models than to persuade them that there are. To see this, suppose we change the questions of this and the last chapter.

Instead of asking the seemingly neutral question that in turn seems to call for a factual response, "What are the expected tonnages in steel for the U.S. versus Japan in the year 2000?," suppose we had asked instead a much more volatile question such as, "Suppose someone very dear and close to you and in their early teens had been raped brutally; whom would you appoint as a *panel of experts* to make the critical decision whether to grant an abortion or not?" Further, instead of asking for a model on something so prosaic as leadership as we did in this chapter, suppose that we had asked instead, "Build a model to make the decision whether to grant an abortion or not?" It comes as neither a great shock nor a surprise that these questions are treated very differently and evince very different responses. Now different assumptions become extremely vital. The discussion becomes even more heated between individuals and groups. The consensus over experts or models that may have flowed freely and easily before has all but evaporated. Everything has suddenly become contentious, as well it should. The problems or questions are no longer well structured. The very phrasings of the initial questions, which were not in dispute before and perhaps were even irrelevant, now become exceedingly critical. The ways in which the questions are posed and the assumptions made bear heavily on what counts as answers. The feelings aroused are so strong that they spill over to the supposedly more neutral and well structured issues, so that if we ask the question of forecasting steel production in the year 2000 and the selection of a president for Zenith Life *after* we have asked the more inflammatory questions, then these have become ill-structured issues as well.

Let us now turn to a more precise description of the characteristics of Analytic-Deductive IS's.

Analytic-Deductive IS's

The inputs into an Analytic-Deductive IS are simple ideas or basic propositions that break a complex phenomenon (e.g., leadership) down into its basic components. The operator is typically a set of mathematical operations that take numerical scores on each of the components and combine them (through either addition, subtraction, multiplication, division, or more advanced mathematical operations, for instance, from calculus) into a single final output (a single numerical score) that expresses the best choice for the problem at hand.

The guarantor is typically composed of two critical parts. Both pertain to the selection of the inputs themselves, although they are operative on all the other phases of the system as well. The first part of the guarantor is concerned with whether one has selected the correct inputs. This part is founded on the notions of "intuitive obviousness" or "self-evidence." For instance, in the case of leadership, it is "intuitively obvious" or "self-evident" what the relevant characteristics of leadership are. The second part consists of the application of one of the most fundamental laws of logic, the law of contradiction. In symbols, this "law" is expressed as "Not (p and not-p)." In words, it reads as follows:

**It is not the case (i.e., it is false) that a proposition
or a statement p and its negation not-p
are both true at the same time.**

It cannot be the case that something is both white and not white at the same time. Or, it cannot be the case that it is both raining and not raining at the same time and in the very same place.

In other words, the Analytic IS places extreme emphasis on logical consistency. Anything or anyone violating the notion of logical consistency, however it is expressed, is to be ignored or dismissed in the strongest terms.

The law of contradiction functions as a kind of screen or test to supplement the criterion of intuitive obviousness in selecting candidates for inputs. For example, if one attribute of leadership, call it p, is more plausible than its opposite, not-p, then since both p and not-p cannot be true at the same time, then one or the other must enter into the proposed model under construction or investigation. More generally still, the science of logic is taken as the supreme guarantor of thought itself. Because logic is regarded as the most basic of all disciplines—every discipline and subject matter has to presuppose the prior existence of logic, or of correct reasoning, in order to reason correctly—logic is regarded as the fundamental guarantor of everything that humans do.

In the sixteenth and seventeenth centuries the guarantor was even more formidable. God Itself was the supreme Guarantor. This among many reasons was why minds of the stature of a Descartes and a Leibnitz were concerned with proving the existence of God. If God existed, and further if God was not evil, malicious, or deceiving, then no matter what the subject one was examining, if one meditated or focused clearly on God, then one would not be led astray in one's reasoning. Thus, if one could establish the truth of at least one central proposition—the existence of God—then one could use this as the base to establish all others.

The great minds of the sixteenth and seventeenth centuries were thus anything but religious fanatics. Instead, they were concerned with the ultimate guarantor of knowledge in general. Or, if one prefers, they were concerned with G.O.D.—the Guarantor of Our Decisions.

At this point it should be—can we say it?—"rather obvious," although perhaps not intuitively so, what the theory of education is that underlies this approach. Education is to be grounded ultimately in teaching logic and first principles because of the inherent weakness, faultiness, and deception in all facts and observations. Was it not, for example, a "fact" at one time that the earth was flat?

Likewise, something is regarded as "objective" in this system if it can be clearly shown that it follows with deductive certainty (i.e., by mathematical or logical proof) from a series of initial propositions that are themselves beyond reproach. Furthermore, since truth is singular, Analytic-Deductive IS's typically, but not always, present one model and one model alone of a problem and accordingly derive one best outcome from that model.

Conclusions

1. There is no doubt whatsoever of the great success of the Inductive-Consensual and the Analytic-Deductive approaches in the natural sciences. Any one of countless models from Newton to Einstein attest to their overwhelming achievements. In the social sciences and in human affairs in general, however, far less "agreement" exists regarding the success of this approach. Even the most seemingly elementary acts of humans possess a "messiness" that renders them very different from the natural sciences. For instance, if a person prefers *A* to *B* (whatever *A* and *B* stand for) and if that same person also prefers *B* to *C*, then a seemingly straightforward axiom of human choice is that the person ought to prefer *A* to *C*. Unfortunately this simple schema does not capture how humans actually behave. C. West Churchman has demonstrated how trivial it is to show why this simpleminded approach to the modeling of human behavior fails miserably:

> It occurred to me that the whole absurdity of preference orderings can easily be shown. . . . There is a time of the day when I prefer whiskey to water. At the same time I often say to myself, "West, I'd prefer that you didn't prefer whiskey to water." It seems altogether reasonable that I talk to myself in this manner. At times I get very angry with the way other drivers behave on the highway. But, when I react foolishly to their stupid maneuvers, I say to myself, "Quit that!" So, when I tell this investigator that I prefer Brand A to Brand B, which "I" is talking? Of course, a Freudian might tell me that this was a conversation between ego and superego. But it need not be. There have been many occasions when I've wished I liked raisins, but I've never felt any moral overtones in this wish. Suppose now that X says, "I prefer A to B, but I wish I didn't; and I prefer B to C, and I'm glad I do." Would you like to infer something about this person's preference for A over C without any more information?[6]

2. A good part of the reason—dare we say "fundamental"—why so many attempts to model human behavior in the form of Analytic-Deductive IS's have failed is that the very first part of the modeling effort is erroneous. Human beings do not start their inquiries into important social problems as geometry does with simple, clear intuitively obvious or self-evident propositions. Instead, they begin with inputs that are already complex. In Russell Ackoff's terms, the starting points for all human inquiries are "messes."[7] *By "mess" Ackoff means that every human*

---

[6]C. West Churchman, *Thought and Wisdom* (Seaside, CA: Intersystems: 1982), pp. 44–45.
[7]Russell L. Ackoff, *Creating the Corporate Future: Plan or Be Planned for* (New York: John Wiley & Sons, 1981).

problem is associated and involved inextricably with every other human problem. For instance, go define and try to solve the "traffic problem" in New York independently of the "crime," "drug," "real estate," or thousands of other problems. The strategy of breaking a complex problem down into its separate parts doesn't hold for human problems of any significance. Thus, the inputs into any process of human problem solving are indeed "messes," where the term "mess" means a *system of interacting problems.* Since there is no prior word in the English language for this situation, Ackoff had to coin one.

3. At the same time, there is also "no doubt" that in the past it has been supremely important for students and practitioners in all cultures to learn how to construct and derive solutions from Analytic-Deductive models of important problems. We have little doubt that this skill will remain an important one in the future as well. However, we believe even more strongly that the problems we face, and will face, demand abilities that go beyond the two IS's we have presented in the last two chapters.

4. Those who are familiar with Descartes' writings know the tortured and involved lengths he went to in order to establish (a) the existence of God and (b) the nature of God's properties. According to Descartes, if one focused one's mind on clear and distinct ideas, one would not be deceived or led astray in one's thinking. However, instead of avoiding error by this procedure, Descartes' version of God led him precisely into error.

Descartes was deceived by his own ideas regarding deception. Descartes believed that if one started with simple ideas and clear, precise definitions of problems, one would not be led astray in solving them. Today, however, the exact opposite seems to be true. Unless we start our investigations of complex problems with a "clear recognition" of their "messiness," that is, their inherent ambiguity and uncertainty, then we seem destined to misperceive the exact nature of the problem. If we have to have precise definitions of complex problems before we can proceed, and if in order to obtain such precise definitions we need to base them on the adoption of a single scientific discipline or profession, then *precision and clarity may lead us deeper into deception and not rescue us from it. By selecting a single scientific discipline or profession, we cut off innumerable other pathways that we could have chosen to explore the nature of our problem.* In this sense, precision and clarity are too high a price to pay for reaching solutions to our problems. The conclusion is not therefore that clarity, precision, or G.O.D. are dispensable or unnecessary, but rather that Descartes' notions of clarity, precision, and G.O.D. are not appropriate for the guarantor that we require for the problems of the twenty-first century.

5. We cannot overemphasize too strongly why analytic thinking has exercised such a considerable hold on the Western mind. Time and time again, model builders have been blinded by their early successes with structured-bounded problems. Hence, they were encouraged to apply the tools they had developed and which worked well for structured problems to modeling an ever-increasing range of unstructured or messy social problems such as transportation, crime, and health care:

> One . . . assumption was that the "hard" part of a problem—which could be expressed in mathematical terms—could usefully be isolated from the human and organizational elements which could thus be eliminated from the analysis. Another false assumption was that implementation [of the supposedly "correct solution"] was an entirely separate activity from the analysis itself.[8]

These tendencies have also been reinforced by the general culture of universities. Since the greatest prestige in universities tends to accrue to those who can publish the findings of their research in the most obtuse, usually mathematical, language, over time the "best" research has become that which can be understood by the fewest number of one's colleagues. As a result, esoteric knowledge, or communication to the fewest number of persons in one's network, has become valued a hundred times over *exoteric* knowledge or communication to the largest and most diverse numbers of persons in the wider community. Solipsism, or the most extreme emphasis on introverted communication, has been generally prized as a sign of higher knowledge. Narrowness has thus become an unwritten rule of the game. Woe unto those who attempt to communicate to the masses, or the community at large, in direct, simple language in popular journals or try to relate different fields of knowledge to one another.

Lest we be accused of exaggerating, listen to one of the most distinguished members of the economic profession, Lester Thurow, describe the narrowness of economics education:

> The discipline of economics is on its way to becoming a guild. Members of a guild . . . tend to preserve and advance traditional theories rather than try to develop new ways of thinking and doing things to solve new problems. The equilibrium price-auction view of the world is a traditional view with a history as old as that of economics itself: The individual is asserted to be a maximizing consumer or producer within free supply demand markets that establish an equilibrium price for any kind of goods

---

[8]R. Tomlinson and I. Kiss, *Rethinking The Process of Operational Research And Systems Analysis* (Oxford: Pergamon Press, 1984), p. xi.

or service. This is an economics blessed with an intellectual consistency, and one having implications that extend far beyond the realm of conventional economic theory. It is, in short, also a political philosophy, often becoming something approaching a religion.

Price-auction economics is further blessed because it can assume mathematic form—it can work hand-in-glove with calculus. Expression in mathematics imparts to the theory a seeming rigor and internal strength. But that rigor easily degenerates into scholarly rigor mortis, as mathematical facility becomes more important to the profession than a substantive understanding of the economy itself. To express an idea mathematically gives it the illusion of unassailable truth and also makes it utterly incomprehensible to anyone untutored in mathematics. Then, too, young scholars aspiring to the profession are required to demonstrate a technical virtuosity in math before they are even considered eligible. By analogy, once the Confucian scholars of ancient China passed a very complicated set of entrance examinations, they used the same examinations to keep others out. Both then and now, all honor is reserved for those who can explain current events in terms of "The Theory," while anyone trying to develop new theories to explain recent developments is regarded with suspicion at best. In economics today, "The Theory" has become an ideology rather than a set of working hypotheses used to understand the behavior of the economy found in the real world.[9]

6. Finally, we can state some rules of thumb that are helpful in overcoming the myopia often associated with models:

- Seek the obvious, but do everything in your power to challenge and even ridicule it.
- Question *all* constraints. The most limiting constraints in building a model or a representation of a problem are usually imposed not by the problem itself but by the mindset of the problem solver.
- Challenge as many assumptions about the problem and the model as possible. Remember that what seems self-evident to the problem formulator is not always evident to others.
- Question the scope or the definition of a problem or model. Frequently what is omitted from the statement of a problem or model is more critical than what is included.
- Question whether a problem is to be "solved," "resolved," or "dissolved." There are important differences between "solving," "resolving," or "dissolving" a problem. They are not necessarily the same. To "solve" a problem means to produce an exact or optimal solution to it. To "resolve" a problem means to seek a solution that is "good enough." On the other hand, to "dissolve" a problem is to realize that

[9]Lester C. Thurow, *Dangerous Currents, The State of Economics* (New York: Random House, 1983), pp. xviii–xix.

there may be some other problem that is more important to focus one's attention on. The old or initial problem may still exist but may not be as important in the broader scope of things.

- Finally, question logic itself. Being logical and being right are not always the same. The more logical a solution to a complex problem sounds, the more strongly it deserves to be challenged.

7. The discussion in this chapter in no way exhausts—for it was not meant to—the extreme and wide variety of different kinds of Analytic-Deductive IS's or more generally "rationalist" systems that range from scientism to mysticism. In the former, one's first principles of truth or knowledge derive from science; in the latter, from revelation or divine inspiration. For how such systems manifest themselves in the social and organizational realm, particularly how they both inhibit as well as facilitate large-scale organizational change, the reader is referred to McWhinney.[10]

We believe that it is no exaggeration to say that, up to the present, the two IS's we have discussed in this and the last chapter have constituted the overwhelming basis for education that most educated people in Western societies have received. The unparalleled challenges we are facing now make it absolutely necessary for us to learn new modes of learning, the task we now turn to in the remaining chapters.

[10]Will McWhinney, *Paths of Change, Strategic Choices for Organizations and Society* (Los Angeles: Sage Publications, 1992).

PART II

COMPLEX THINKING

CHAPTER
4

Multiple Realities:
The Third Way of Knowing

BIGOT, n. One who is obstinately and zealously attached to an opinion that you do not entertain.

EDUCATION, n. That which discloses to the wise and disguises from the foolish their lack of understanding.

PREJUDICE, n. A vagrant opinion without visible means of support.
Ambrose Bierce, *The Devil's Dictionary* (New York: Castle Books, 1967)

For most of the past four decades, the CIA has been badly overestimating the performance—and prospects—of the Soviet economy.

The reasons may prove critical lessons for the future. In a nutshell, critics say, the CIA was led astray by the American bent for high technology: It relied too heavily on computer models and not enough on old-fashioned, first-hand observation and reporting.

In the early 1960s for example, the CIA said the Soviet economy was already half as big as America's and growing more than twice as half. In reality, the Soviets were beginning the decline that has led them to the brink of collapse.

"Common sense was just not applied," a senior defense official says.

By contrast many of America's allies trusted more to first-hand observation and their experts' own judgments. On that basis, analysts in Sweden and Britain have long regarded the Soviet Union as essentially a third-world economy—an "Upper Volta with missiles." But their views were dismissed in Washington.

The idea of using econometric models for intelligence estimates—which parallelled the general trend in economics at the time—was the brainchild of Harvard University's premier Soviet economist, Abram Bergson, who touted the technique as more precise than the old system of sending in agents.

The trouble was that using computer models forced the CIA to rely on Soviet statistics, which were known to be flawed—or sometimes even fabricated.

Intelligence analysts tried to factor out the flaws in Moscow's data, but the Soviet statistics were so inflated that they were almost useless.

Robert C. Toth, "Espionage/Spying by the Numbers, CIA Misread Soviet Economy, Agency's Dependency on Computer Models—Instead of Observation—Produced Results That Were Widely Askew"
Los Angeles Times, Friday, August 3, 1990, p. A5.

The academic/professional mind is a marvelous thing to behold. Because of long and arduous years involved in mastering a particular discipline, the academic/professional mind easily becomes the prisoner of a particular way of viewing the world, or, in the terms of this book, a particular IS. For this reason, crossing academic disciplines or professional boundaries is a harrowing experience. It is worse than crossing foreign cultures—it constitutes culture shock of the highest order. Many of the taken-for-granted, self-evident propositions of the members of a particular academic discipline or professional group—which of course are intuitively obvious and make good sense to them, and hence, are accepted without much debate—appear absolutely ridiculous and absurd to those of another. While those on the inside and outside appear to be using English, they might as well be speaking in different tongues. The differences in connotations are vast because of the dissimilar pictures the proponents have of the world and of the vast differences in assumptions they entertain.

The personal or public examples that we could use to illustrate these points are so numerous that it's difficult to pick one. However, Mitroff recently had an experience that is as good as any. Approximately sixty-five scholars convened in the beautiful wine country of Northern California to share insights and ideally advance the still struggling discipline/profession of strategic management. In general, strategic management is concerned with putting the study of business strategy or strategic planning on a scientific basis. Some of the more detailed concerns of this fledgling field are: which business strategies work for which firms under which economic and social conditions; do managers really make a difference in the economic performance of a firm or do impersonal, external economic factors contribute more?

Mitroff was the moderator of a panel concerned with process. All the panel members, including Mitroff himself, were in general agreement—and hence constituted a mini Inductive-Consensual IS—with the notion that the decision, interpersonal, and managerial processes that the members of a firm used to govern themselves had an important bearing on a firm's economic competitiveness. For this group, it was obvious that the Japanese, Koreans, and West Germans were successful because of key differences in the internal management processes they used to produce goods and services. For instance, in Marysville, Ohio, where Hondas are built by American workers under the supervision of Japanese management methods, the cars produced are the same high quality as those produced in Japan. Many of the cars made in Ohio are shipped back to Japan for purchase by the Japanese themselves.

The audience was ideologically split. For the most part, the economists held to the view that process was not as important, or, if it was, they wanted the hard numbers to prove it. For the first group, process mattered not merely because it was obvious to them, but because the kinds of professional experiences they had predisposed them to this conclusion. Through individual case studies and even personal consultation, the first group got inside of individual firms and knew them intimately. In addition, they studied the grand differences between the managerial practices of companies in different countries. Some of these studies were founded on data,[1] to be sure, but not the kind needed by the economists before they would be convinced.

In short, the first group studied firms from the inside out. The economists, on the other hand, studied firms from the outside in, if indeed they ever got "in." As economists, their fundamental unit of analysis (i.e., the inputs into their collective IS) was the firm or group of firms. Rarely, if ever, did they dip below the firm as the fundamental unit of analysis to penetrate the "mysterious black box" that constituted an organization. In other words, they studied the inputs into firms and the resultant economic outputs from them. Rarely did they get inside a firm to study what went on to transform inputs into outputs. Their whole mode of training predisposed them not to do this. For economists, the external forces or factors of production acting on firms—for instance, the size and power of competition, economic and financial markets— were the important factors determining the behavior of firms. No wonder there was an almost unbridgeable gap between the two groups.

To illustrate this even further, consider the following. During the session one of the economists in the audience—although of an unusual sort since this person was one of the very few equally conversant in the language of behavioral science and of economics and hence did dip inside firms to study them—asked the panel, "Can you conceive of situations where process doesn't matter or is not as important as external factors?" The members of the panel either couldn't or didn't want to respond to this question at the time. After the session was over, Mitroff approached the individual who raised the question—a personal friend and one whose opinions he respected—and said, "Tom (not his real name), think about how your professional mentor (an internationally famous systems scientist) would have responded to your question? He would have asked you another question in turn, 'Imagine a subject in a closed room performing a simple psychology experiment, for instance,

---

[1]For two typical and very good examples of this approach, see James C. Abegglen and George Stalk, Jr., *Kaisha, The Japanese Corporation* (New York: Basic Books, 1985); and Peter Drucker, *The New Realities* (New York: Harper & Rowe, 1989).

tapping the point of a pencil back and forth between two squares, the object being to get as many correct hits in a specified amount of time. Now, the question is, does respiration or breathing matter in this experiment? Of course it does! It's always there. It's the stuff in the background that's taken for granted. Asking how much it contributes to the experiment is nonsensical. You can ask whether some forms of breathing are more effective or not, but you can't ask whether humans as we know them never need to breathe. Wouldn't we regard someone who even asked the question whether respiration mattered as strange, perverse, and perhaps not even sane? Isn't the same true of 'process'?"

Conferences like the one Mitroff attended cannot be described in purely rational terms. Sitting there, Mitroff was reminded of a passage in one of John Dewey's books. Dewey described a simple world with merely two entities: (1) eaters, that is, things that ate other things in the environment, and (2) the things that were eaten. Knowing the academic mind as well as he did, Dewey conjectured that academic disciplines and professional specialties would arise that would focus on each of these two entities to the exclusion of the other. Two camps would immediately spring into being, each arguing that its entity was the more fundamental. Naturally, each camp would also argue that its favored entity could be studied without reference to the other. What nonsense! Since each entity depended fundamentally on the other for its existence, neither could be studied nor understood completely without significant reference to the other. The same pertains to both of the sides or camps in the conference that Mitroff attended.

Whether the reader accepts fully the position of either group is not the issue. The real point is that in general each of the various academic disciplines and professions constitutes different mixtures of the previous two IS's or ways of knowing that we encountered in Chapters 2 and 3. This realization is the starting or take-off point for the next way of knowing that we describe in this chapter. However, we must first introduce some background concepts.

We also need to know that exceedingly practical considerations turn on the discussion. For instance, academics may have the luxury of participating in the kind of standoff in which Mitroff was embroiled, but today's managers and executives do not. A lament all too commonly and frequently heard in today's organization is:

> I can't tell you the incredible talent we've got around here. If we could only "get it together," we could beat the pants off the Germans and the Japanese. The thing that depresses all of us is that we really want to agree. We just can't seem to achieve it. We come from such different educational and professional backgrounds that even when we seem to be using

the same words, they mean entirely different things to each of us. The trouble is that we have no choice but to find agreement between us. The engineering, marketing, planning, production, and sales teams have to be involved *from day one* in the design of critical products. We can't afford to have engineering produce designs any longer that customers don't want, and we can't have marketing and sales force through products that are impossible to manufacture economically. We're divided by our differences. Instead of our differences being one of our greatest strengths, they are one of our greatest liabilities. Is there a super model out there somewhere that could integrate our differences into a common framework that we could get behind?

The Computer as a Kantian IS

The IS that is the subject of this chapter is considerably more complex than those of Chapters 2 and 3. It is in fact a special and complex "mixture" of both. It combines the "model part" of Analysis and the "data part" of Agreement into an interactive whole. To see how and why, we need to understand something of the more prominent features of the philosophy of Immanuel Kant. Kant's ideas form the general background framework for the present IS.

Ever since Kant, educated people have realized that both the experience of reality as well as its description are heavily dependent on the structure of our minds, much more so then empiricists would have us believe. Contrary to the common-sense notion that reality is "something out there" uninfluenced by human minds, we humans contribute a great deal of our nature to what we experience as reality and how we describe it.

One of the best ways to see this is to consider a simple analogy: the human mind as a computer. This way of looking at the mind is so powerful that we are convinced that were he alive today, Kant himself would have used it as a vivid way to demonstrate his theory of human knowledge.

Computers are carefully designed to accept inputs from the outside world that are of a very special kind. The elements fed into a computer must be of the right form or the computer will not be able to recognize them. Most computers accept either electrical signals or cards with holes punched in them. In the case of electrical signals, either magnetized tapes or keys at a computer keyboard are used to enter the inputs into a computer.[2] With punch cards, holes are either punched in cards

[2]The latest computers now allow for voice and handwritten inputs. However, since these are typically converted into electrical signals, the same principles apply.

or left filled in. In both cases, the computer is able to read input data as a series of "ones" (if a "one" is present, then a portion of a tape will be magnetized, a key at a computer keyboard will be pressed, or a punch hole will be made in a card) or "zeroes" (in the case of a magnetic tape, a portion of the tape will not be magnetized, or a computer key will not be pressed, or in the case of a punch card, the hole will be left filled in). Computers are specially constructed to receive only very certain kinds of input data—ones or zeroes—that are imprinted on valid media, tapes or cards carefully manufactured to be of certain sizes. These in turn must be input into the correct receiving mechanism or hardware attached to the computer.

One could literally throw cards or tapes at a computer all day long, or send electrical signals representing so-called "hard data" from the external world to a computer, but unless it was carefully designed and built in such a way to accept tapes, cards, or electrical signals (if it had the right kind of hardware to read incoming tapes, cards, or electrical signals), then the computer would not be able to have experience of the outside world in the form of zeroes and ones. Further, if it were not for the internal programs or software stored in the computer capable of "making sense" of zeroes and ones, then the computer would never be able to recognize patterns in the stream of zeroes and ones and hence reach conclusions about the nature of the outside world. Thus, at least two things are needed to have experience: an ability (1) to receive input data from the outside world, and, once in, (2) to make sense of what has been received. *Neither of these two critical abilities is a property of external things themselves.* They are properties of the computer, and, by analogy, our minds. The structure of our minds thus plays a fundamental role in: (1) what we experience as reality, (2) how we experience it, (3) what we characterize as reality, and (4) how we characterize it.

The pattern recognition routines as well as the input mechanisms of computers do not determine the *content* of the input, but they do prescribe its *form.* Computers themselves set severe constraints on the form of what is fed into them. The patterns that are fed in supposedly encompass data descriptive of some problem in the outside world; for example, the incidence of various crimes in a certain neighborhood, the amount of air pollution over a city, and so forth. Data that are not input in this form cannot be recognized by the computer.

Consider another, even simpler example. Imagine the existence of a highly intelligent wine glass, one with a reflective mind that was able to ask itself questions. Suppose such a wine glass asked, "How is it that no matter what the quality, bouquet, age, or origin of the wine that is poured into me, it always has the same shape? How is this possible?" If our wine glass were intelligent enough, it would come to the realization that *it* supplied some critical feature of its experience of the wine that was poured into it, namely, its shape. Everything else that was characteristic of the wine, its flavor, color, bouquet, age, country of origin, was supplied by something or someone else external to the wine glass. None of these things were determined by the wine glass.

For human beings to have experience or gain knowledge about the external world, something must be built into the internal structure of their minds that is capable of receiving data or facts characteristic of the outside world or presumably reality itself. The computer is not able to obtain certain facts in the form of zeroes and ones about the outside world unless the data are input in the right form. Similarly, the wine glass is not able to receive data in the form of wine from the outside world unless it is made in a certain kind of shape and material so that it can contain the wine.

Knowledge about the world is a result of an *interaction* between both the structure of the computer and that of the data that can be input to it, between the shape and the material of the wine glass and the viscosity of the wine. Both the computer and the wine glass supply a critical, crucial ingredient to both the structure of reality as well as to our knowledge about it. The wine glass does not create the external reality, the wine, that is fed into it; it merely possesses a shape and a material whereby wine can be stored or contained within it. Likewise, computers do not create what is on the cards, tapes, or electrical signals fed into them, and in this sense they do not create the outside world, but they do possess a set of internal *categories* so that whatever external reality is fed into them can be pigeonholed.

In the case of computers, these categories are in the form of "addresses." They work much like mailboxes along a street. They tell a computer where to put the series of zeroes and ones that are punched on cards, tapes, or embedded in signals. For instance, computers are typically built to read the separate columns on each card and to place the zeroes and ones in each column into a distinct mailbox or address inside the computer's memory. Once all the data from the cards are inside the computer, the computer's internal programs take over. The computers in essence may be programmed to add the zeroes and ones on all odd (or even) mailboxes together.

The Multiple Realities Approach as a Kantian IS

The various professions and disciplines put special boundaries around the realms of experience or expertise they treat. In terms of our computer analogy, the various professions in effect have different computers. They take in very different inputs and process them differently. In terms of the wine glass analogy, the professions not only pour and cast their problems into different wine glasses, but they also take in altogether different wines. In other words, the various professions treat the "same" problems differently as well as focus on or receive different problems to begin with. In effect, the boundaries between professions are the differences between the categories they use to structure the problems they treat.

Consider, for instance, the problem of drug use and addiction, which is alarmingly high in the United States (the United States consumes up to 80% of the world's cocaine supply). The number of different lenses through which this problem can be viewed, while not unlimited, is nonetheless large. For instance, in principle, each of the following disciplines or professions has a very different perspective on the problem; as a result, each uses different variables to structure or represent the "problem," and consequently collects very different kinds of data: education, economics, social work, medicine, criminal justice, psychology. In educational terms, the problem is one of educating young people and their families to the dangers of drug use; for instance, a typical TV ad is: "This is what happens to your brain when you use drugs; it becomes fried like an egg." In the language of economics, the problem is the huge profits associated with the production and consumption of illegal substances; take the profit or economics out of the picture, and you'll dramatically lower the crimes associated with it. In the language of social work, the problem is the breakdown of the family, the lack of male role models, and so on. In medical terms, the problem is one of treating the physiology of drug addiction. For the criminal justice system, the problem is more cops and money for policing. Finally, for psychology, the problem is the despair of people in inner cities and the associated problems of low self-esteem that is partly responsible for their joining gangs, engaging in self-destructive behavior, and other factors.

Even these hardly begin to exhaust the other filters we could use to define the "drug" problem and formulate policy options or solutions. The problem of homelessness provides another example:

> The "homeless problem," in the United States . . . is not a problem of inadequate housing alone, but of several interlinked problems—alcoholism, drug abuse, unemployment, mental illness, high land prices. Each is the concern of a different bureaucracy, none of which can deal effectively with the problem on its own, and none of which wants to cede its budget, authority, or jurisdiction to another. *It is not merely the people who are homeless, but the problem* [italics ours].

Drug abuse, too, requires integrated action by many bureaucracies simultaneously: police, health authorities, the schools, the foreign ministry, banking, transportation, and more. But getting all these stacked effectively in concert is almost impossible.

Today's high-speed technological and social changes generate precisely this kind of "cross-cutting" problem. More and more of them wind up in limbo, and more turf wars break out to consume government resources and delay action.[3]

McWhinney puts it as follows:

"Homelessness" is not a well-bounded problem; attempts to solve any aspect of it spill over into other areas of social concern. A free meal regularly offered to . . . street folk in Los Angeles drew increasingly large crowds that compounded the difficulty locally and created other dislocations at least as grievous as the one on which the meals program had focussed. Applying a local or partial solution will typically produce greater difficulties in the surrounding environment. We need to treat the unboundable, supra-problems differently. They are issues, not problems.

An *issue* is an unbounded, ill-defined, overwhelming complex of problems. Homelessness is, in the minds of most Americans, an *issue*. The word "issue" itself expresses a sense of outflowing, uncontainable at a point in time and space. To dam it up in one place is to invite overflow into another. If attempts to resolve an issue are not to create more problems than are solved, an issue must be approached as a whole with strategic awareness of capabilities, recognizing that neither the content of the issue nor the process of resolution can be contained by well-bounded constructs.[4]

To draw out deeper aspects of this approach, we will cast it as before in the form of an IS.

The Multiple Realities IS

It is more difficult to separate the various components of the Multiple Realities IS than the two we have discussed in the previous chapters. Indeed, this is a prime feature of this particular IS. *In what we call the Multiple Realities IS, data, facts, or observations are not separable in principle from the theory or the model that we construct of a problem.* The data, facts, or observations one collects about a problem are highly dependent on the theory or model one has of it. In other words, the prior model that

[3]Alvin Toffler, *Powershift* (New York: Bantam Books, 1990), pp. 262–263.

[4]Will McWhinney, *Paths of Change, Strategic Choices for Organizations and Society* (Los Angeles: Sage Publications, 1992), pp. 82–83.

one has of the world determines subsequently the data one collects from it. Taken together, model and data form an *inseparable whole.* A particular model/data combination or coupling defines a view of reality, or a "reality" for short.

Think back to the model of leadership we considered in the last chapter. Depending on the model we adopt, we could in principle collect very different data on each of the four candidates. Or if the data were already given to us by someone else—as they were in the initial *Harvard Business Review* case by someone else using their own implied model of leadership—then the data would be selectively reinterpreted to fit with a particular model. Like our computer analogy, unless the data are in the special form that a model treats, then they cannot be fed into the model and, hence, recognized by it, let alone operated on.

To repeat, data and model/theory are not separable in this or in any other system. The data one collects from the world are a strong function of the images, models, and/or theories we have of it. Data, in other words, are not self-organizing entities as Inductive-Consensual ISs assume. The pattern or patterns we find in data are put there by us—they involve our images, models, and/or theories of the world. Since, in principle, different disciplines and professions do not share in toto the same images, the data they collect on important problems are also different. The recognition and appreciation of this constitute one of the main features of the third way of knowing, the Multiple Realities IS.

As a consequence, inputs into the Multiple Realities IS are considerably "more complex"[5] than those of either the Inductive-Consensual or Analytic-Deductive IS's. The input or inputs into a Multiple Realities IS are composed of two strongly interrelated parts: (1) a *data set* that is distinctively *coupled* to a particular image or model/theory of the problem, and (2) a *range* of different data-model/theory *couplings* that represent various views or representations of the problem.

One of the most important ways in which Multiple Realities IS's differ from Inductive-Consensual and Analytic-Deductive IS's is that Multiple Realities IS's do not assume that there is one and only one way to define important problems. Unlike Inductive-Consensual and Analytic-Deductive IS's, the assumption is not made that the definition

---

[5]It may be fairer to say that Multiple Realities IS's embody a different kind of complexity than Analytic-Deductive IS's. We do not deny for one moment that Analytic-Deductive IS's can utilize extremely large and complex models and their inputs can also be extremely large and complex. However, Multiple Realities IS's embody a different kind of complexity—input complexity. This differs from model complexity by which we typically mean a large number of complex and intricate mathematical operations used, to transform inputs into outputs. In Multiple Realities IS's, the input is already confounded with some kind of a model.

of a problem is unproblematic. Instead, the assumption is that on problems of any significance, the analyst, decision-maker, or policy-maker needs to see explicitly a *range* of different representations of the problem so that he or she can *participate actively* in the problem-solving process and not merely be a static recipient of the end results of the Inductive-Consensual or Analytic-Deductive processes. In the Inductive-Consensual and Analytic-Deductive IS's, the decision-maker is usually presented with a single, end outcome that the individual either accepts or rejects with the additional option of initiating a new inquiry.

The key phrase and concept in the preceding paragraph is the *active participation* of the decision-maker in the inquiry process. In the Multiple Realities IS, the decision-maker is a substantial part of the operator. The decision-maker must view the range of representations of his, her, or society's problems and decide which one, if any, is "best" or applies to the particular situation at hand. Since one of the fundamental purposes of this system is to aid the decision-maker in forming a distinctive model, it is "hoped" as result of viewing the range of models that the decision-maker will be in a stronger position to produce, or at the very least, participate in the production of a new model that is a creative synthesis of the initial ones.

Unlike the previous cases of the Inductive-Consensual and Analytic-Deductive IS's that we encountered, the operator cannot be purely formalized or mathematized in this system. It is thus highly dependent on the personality and creative abilities of the particular decision-maker. Instead of regarding variety as a strength before making a choice in complex situations, those whose personalities prefer a single, clear-cut option are more likely to feel overwhelmed and confused by a Multiple Realities IS.

While the output of a Multiple Realities IS can be the choice or selection of a single model that best synthesizes the variables of the separate input models, it is also just as likely to be the demand that the initial range of models be expanded before one choses. This brings into focus another critical aspect of the system's input/operator—the "executive." The executive is the entity or person who picks the range and particular kinds of initial models to present to the decision-maker. In general, there are no formal rules, or at least none that we are aware of, to guide the executive in his or her selection in all cases. One can, however, state some heuristics or rules of thumb that are helpful.

On any problem of significance, a minimum requirement is that the models *not* be drawn from any single discipline but from a range of them. Thus, in the drug example, one does not want to develop all of one's models from, say, medicine, although there could well be two different models from a field representing diverse viewpoints. In general, one tries to present a range of models that cover as broadly as possible the concerns inherent in the situation that one is explicitly considering in any important decision. Thus, an economic model of some sorts will more than likely be chosen since financial resources play a prominent role in most decisions, almost by definition. Since technology also has a key role, some scientific/technical model or models may also be selected. Since one cannot in general decouple psychological and sociological concerns from important decisions, at least one psychological and/or social model will also be present.

The reason the executive cannot be purely formalized is that it is highly dependent on the exercise of wisdom. And, above all, wisdom is the one factor that cannot be cast into a mathematical formula or procedure.[6] If wisdom is not present, then the system will likely degenerate into a monumental exercise in nonsense. This is precisely why the Inductive-Consensual and Analytic-Deductive IS's have been so persistent and perennial in human affairs.

Systems do not last for long unless they have something strong going for them. While the Inductive-Consensual and Analytic-Deductive IS's may disagree strongly over the starting and ending points of inquiry, plus what is an acceptable guarantor for knowledge, they do agree on one important element: The starting and ending points of knowledge must be simple, clear, and unambiguous. The Inductive-Consensual IS locates the simple starting points in observations or agreements; the Analytic-Deductive locates them in simple notions or propositions that supposedly are self-evident to all. Both IS's offer comfort and security in the thought that the world is basically simple and stable.

Unfortunately, if the history of philosophy has demonstrated anything, it is that there are no simple, certain starting points for anything. The argument between the Inductive-Consensual and Analytic-Deductive IS's is over what the appropriate "simple starting points" are, not whether they exist. Yet, if philosophy has shown the lack of these starting points, that does not mean humans have accepted this conclusion gladly.

In one of his most important books, *The Quest for Certainty*,[7] the eminent American philosopher John Dewey argued that the history of Western philosophy was the quest for certainty, and, more fundamentally, a neurotic one at that. Given the complex, uncertain, and perilous world in which humans live, the reason for such a quest, however neurotic it may be, is nonetheless understandable. Yet this quest does not

[6]C. West Churchman, *Thought and Wisdom* (Seaside, CA: Intersystems Press, 1982).
[7]John Dewey, *The Quest for Certainty* (Carbondale: Southern Illinois University Press, 1929).

contribute positively in the long run to humankind's increased knowledge of the world. *Once one abandons the playing field of certainty, one has no choice but to face the fact that uncertainty itself is a fundamental characteristic of the process of knowing and even of knowledge itself.*

What then is the guarantor of the Multiple Realities IS? This guarantor consists primarily of two major entities: (1) the *range* of views itself, and (2) *hope.* Indeed, the guarantor is largely contained in the "hope"— and note that "hope" is implicit in the guarantor of *every* system; the current system only makes the role and the necessity of hope more apparent—that as the result of viewing explicitly the range of different views, the decision-maker will thereby achieve a deeper understanding of his or her own problem. Embedded in the guarantor is the notion that on important problems, the ability to see alternative views or models/data couplings of it is not a luxury but rather a fundamental necessity. *The production of multiple views of a problem is an explicit requirement for inquiry and for knowledge itself.*

That the decision-maker plays such a strong and fundamental role helps explain why Multiple Realities IS's are often called Interpretive Systems. Multiple Realities IS's are highly dependent on the ability of (and hence, presuppose the prior presence of) a human to interpret the range of initial views, to decide which if any are best for the problem, and to synthesize a new view from those presented initially. A complex human equation is thus inherent in the Multiple Realities IS: no humans, no interpretations; further, since no interpretation is ever beyond all challenge, no interpretations, no risks; no risks, no knowledge. *In short, knowledge is inherently a very tricky, risky business.* Remove all risks and one not only removes all knowledge but its very possibility. Knowledge is not for the weak willed or faint hearted—safety and knowledge are actually enemies.[8]

The poet Robert Bly puts the above point even more forcefully:

Teachers and therapists often have a strong Cook, Mythologist or Magician inside. But if a teacher has not developed the Wild Man or Wild Woman, that person becomes the strange being we call an "academic" whose love of standards is admirable in every way, but who somehow filters the wildness out of Thoreau or Emily Dickenson or D. H. Lawrence even as he or she teaches them. Not all teachers do that, thank God, but universities shelter a lot of them.[9]

Something is "objective" in this system if and only if it is the result or the product of a range of different views. Conversely, something is

---

[8]Robert Bly, *Iron John A Book About Men* (Menlo Park, CA: Addison-Wesley, 1990), p. 230.
[9]*Ibid.*

not objective if it is the result of one model or one group of experts no matter how strong the model or the agreement between the experts. This is the basis for the assertion we made earlier that there is not one and only one definition or concept of objectivity, and other key terms as well. To be truly educated in this system means one not only possesses the ability to appreciate and understand the different meanings of key terms, but to know which ones are appropriate for which kinds of situations.

In this system, one is not said to be educated unless one can frame important issues from a multitude of different models/data couplings. One must be broadly educated across the different sciences and professions. Thus, from the standpoint of this system, the current system of education—at all levels—is not only a disaster, but unethical and unaesthetic. The partitioning of knowledge into separate, autonomous disciplines as the foundation for the structure of the current university is unethical because it grossly distorts the nature of important issues and problems. In this sense, it thus seriously hampers our ability to frame and formulate solutions for important social problems by distorting their nature. It is also unaesthetic because a reductionistic approach to knowledge distorts the style and the mood within which important problems are discussed.

If both the Inductive-Consensual and Analytic-Deductive IS's speak in the indicative mood—a potential fact either "is or is not" and a potential theory or model is either "true or false"—then Multiple Realities IS's pose their results in the hypothetical or conditional. For instance, "*if* one accepts the background assumptions of this or that particular model, discipline, or group, *then* the representation of a problem 'is' such and such, or, better yet, 'might be or can be considered to be' such and such." The acceptance of a particular model thus makes it possible for us to see the data associated with a particular way of viewing the world.

It is tempting to conclude that the Multiple Realities IS is really only an Analytic-Deductive or Inductive-Consensual IS in disguise. After all, isn't the final goal of the Multiple Realities IS to achieve a better single model of a problem? Doesn't the Multiple Realities IS merely prolong the process by having us first view a number of incomplete models? Merely to ask such questions is to miss the whole point and spirit of this system.

Implicit in its way of viewing the world is the strong assumption that there may well be no finite set of models—certainly not a single one— that can ever cover or capture all aspects of reality. The essential purpose of the Multiple Realities IS is not to come to a single model that is necessarily true. Its main purpose is to allow a decision-maker to act.

The decision-maker in this system is not always a scientist interested in Truth per se.

The model or position that a decision-maker adopts for action is provisional on every account, although the language of politics often forces most decision-makers to sound far more convinced or dogmatic than wisdom would allow them. It is not the truth of the model or view that the decision-maker forms that allows the individual to act—the model forms a blueprint for action. Through action, the decision-maker comes to learn of the "working truth" of the model, where "truth" is defined by the ability to respond not only to one's immediate problem, but to anticipate future problems, and thus gain an invaluable edge on working on them before they have gotten out of control.

The fate of nations will be increasingly determined by the ability of its citizens to view problems from multiple professions and multiple disciplines. The distinguished political scientist Alexander George has pinpointed vividly the dire and disastrous consequences that can result when methods appropriate for well-structured problems are applied inappropriately to ill-structured ones that explicitly require multiple views:

1. When the decision-maker and his or her advisors agree too readily on the nature of the problem facing them and on a response to it.
2. When advisors and policy advocates take different positions and debate them before the executive, but their disagreements do not cover the full range of relevant hypotheses and alternative options.
3. When there is no advocate for an unpopular policy option.
4. When advisors to the executive thrash out their disagreements over policy without the executive's knowledge and confront him or her with a unanimous recommendation.
5. When advisors agree privately among themselves that the executive ought to face up to a difficult decision, but no one is willing to alert him or her to the need to do so.
6. When the executive, faced with an important decision, is dependent on a single channel of information.
7. When the key assumptions and premises of a plan that the executive is asked to adopt have been evaluated only by advocates of that option.
8. When the executive asks advisors for their opinions on a preferred course of action but does not request a qualified group to examine more carefully the negative judgment offered by one or more advisors.
9. When the executive is impressed by the consensus among his or her advisors on behalf of a particular option but fails to ascertain how firm the consensus is, how it was achieved, and whether it is justified.[10]

---

[10]Alexander George, *Presidential Decision Making in Foreign Policy: The Effective Use of Information and Advice* (Boulder, CO: Westview, 1980), pp. 23–24.

Because of the enormous role that Inductive-Consensual and Analytic-Deductive thinking have exercised on the Western mind, the way of knowing outlined in this chapter is not easy to grasp. For this reason, we want to show later how it is the historical basis and precursor of the fifth way of knowing, Unbounded Systems Thinking. However, we must first examine conflict, the subject of the next chapter.

CHAPTER
5

Conflict:
The Fourth Way Of Knowing

Contradictions [drive] a person or a community from the first awareness of negation to a resetting on a new truth. Few people have been comfortable living
without some mode of establishing what is true, right, better or more beautiful.
Yet progress can come by letting go of one truth to explore another. This explora-
tion is the heart of the dialectic.
  Will McWhinney, *Paths of Change, Strategic Choices for Organizations and Society*
(Los Angeles: Sage Publications, 1992), p. 213

If in Chapter 3 we were critical of analytical reasoning, it was only the
particular kind presented there. We do not doubt for one moment that
all of us need to be better able to analyze complex problems. In this
sense, the poor ability of students and executives alike to analyze complex problems constitutes one of the main shortcomings of the U.S.'s
current educational system.

The key question, however, is, "What kind of analytical reasoning is
appropriate for today's world?" In our view, the ability to zero in on the
critical assumptions or key premises that underlie complex issues is
desperately needed. Ways of accomplishing this are the subject of this
chapter and Chapter 8. In a deeper sense, how to uncover, challenge,
and replace key assumptions with more appropriate ones is the subject
of this entire book. Notice that it is not an either/or approach. Students
must master the kind of analytic skills that are involved in building
models and deriving solutions from them. However, they also need to
develop the critical ability to challenge the assumptions on which all
models rest.

Suppose in Chapter 2 that instead of eliminating experts whose
judgments of U.S. versus Japanese steel production in the year 2000
were *too far from the average,* we eliminated instead experts whose judgments were *too close to the average.* If we did this, then we would be using
another form of the Delphi known as a Policy-Delphi. Alternatively,

suppose that in Chapter 4 we had picked two (or more) views of the drug issue that were either in complete or the strongest possible opposition to one another—for instance, staunchly pro and con with respect to the legalization of drugs—then we would have a very different IS. This system, the Dialectic, is the topic of this chapter.

While the form of the Dialectic we will discuss obviously derives from the long and rich history of philosophy, it is not strictly a product of it. It is in fact the use of the Dialectic as a practical decision tool, not as an abstract idea, that makes it so very different in this book. The form of the Dialectic we discuss is as a bridge or transition to the kind of new thinking we examine in the next chapter.

As the result of our actually using the Dialectic in a wide variety of complex problems and helping top business and government executives employ it as well, we are convinced that the ability to examine complex situations from a dialectical perspective will be a critical analytic ability for problem-solvers to master.[1]

The Failure of Success: The Case of the U.S. Auto Industry as a Natural Dialectic

The Dialectic IS is best illustrated through several concrete examples that show how critical are the assumptions a business or industry makes about itself. In particular, we want to offer two slightly different treatments of the assumptions behind the strategies of the U.S. automobile industry. The first comes primarily from a number of scholarly books on the general history of the industry.[2] The second is from a study done specifically of General Motors. Both indicate that the life cycle of the automobile industry's critical premises was no more than sixty years at best. From roughly 1910 to 1970, the industry's underlying assumptions were not only valid, but allowed for the overwhelming success of one of the most important industries the world has known. But then almost overnight—in the span of some five to ten years—the American automobile industry virtually collapsed. It became so out of touch with reality it almost went down the tubes—for good.

Table 5.1 lists what we refer to as the "Unwritten Rules of the American Automobile Industry." The first column is our "straight or

[1]See Richard O. Mason and Ian I. Mitroff, *Challenging Strategic Planning Assumptions* (New York, John Wiley & Sons, 1981); see also Ian I. Mitroff, Richard O. Mason, and Vincent Barabba, *The 1980 Census: Policymaking amid Turbulence* (Lexington, MA: Lexington Books, 1983).

[2]See Ian I. Mitroff, *Business Not as Usual, Rethinking Our Individual, Corporate, and Industrial Strategies for Global Competition* (San Francisco: Jossey-Bass, 1987).

Table 5.1 The Unwritten Rules of the American Automobile Industry

Straight Talk	*Professional Talk*
1. An easy, short childhood is the best preparation for adulthood and maturity.	It was a distinct advantage that by about 1930 the modern car industry was firmly established, its competitive practies well understood, its major product design features firmly locked into place, etc.
2. We are stable now and forevermore; the broader world is stable.	The competitive dynamics and basic business of automobile production is essentially stable and well known.
3. "They love us" (i.e., our products); they're loyal, won't switch; we can take them for granted; we can assume consumer stability.	The tastes of U.S. car buyers are standardized and stable; thus, except for yearly styling changes, we do not have to make radical or substantial changes in our product. U.S. car buyers will not demand a new type of car that we have never built in large volume before.
4. Nothing new will be invented that will radically shake up our product; essentially, we know it all; the stability of car technology can be taken for granted.	The design/production of future cars will not require fundamentally new manufacturing processes or technologies.
5. Our focus need not be broader than the driver; a restricted focus of innovation can be assumed.	Innovations relating to driver comfort are more important than fundamental technical innovations in the basic product.
6. Don't change until forced to; resist, deny change; put your major energies into denial and resistance.	We can succeed by not spending money on fundamental innovations until forced to by governmental regulatory agencies, foreign competition, consumers, etc.
7. Get your priorities wrong; innovation can take a backseat to efficiency.	Because of GM's dominant industry strategy (under A.P. Sloan), based on clever marketing to different demographic segments of the population and frequent style changes, technical innovation was subordinated to efficiency in production, i.e., efficiency is more important than innovation.
8. Keep getting your priorities wrong; good labor relations can take a backseat to efficiency.	Efficiency in production is more important than good labor relations; good labor relations are not important to efficiency.

(continued)

Table 5.1 (*Continued*)

Straight Talk	Professional Talk
9. We're so big and powerful, smug and secure that we can shut out the whole world; we can charge and pass on anything we want to our customers. So what if we're arrogant?	Foreign competition will never be significant; therefore, U.S. car makers will not be prevented from passing the higher costs of production necessary to keep up with the competition on to consumers.
10. Since we don't need much innovation, we can finance whatever we want to.	The capital and debt capacity required to finance whatever innovations are required will be readily available.
11. Managers don't need challenge in their work; the restricted focus/nature of managerial work can be assumed.	As the business of car making became well understood, not only did managerial work become routine, but it was desirable that it did so. The challenge of managerial work was not necessary to the long-term success of the industry.
12. If you want to get tunnel vision, then you have to reward it. We are masters at creating a system for producing managerial myopia.	An extremely handsome bonus system that rewards top management for short-term thinking is not hazardous to the long-term interests of the entire industry.
13. Workers don't need challenge in their jobs; the restricted focus/nature of all jobs can be assumed.	Workers are willing to trade money for challenge in their jobs.
14. Keep everyone small-minded and uninvolved.	It is not necessary to engage most employees in the larger purposes of the business.
15. Don't rock the boat, don't bite the hand that feeds you; these rules pertain to the unwritten culture of the industry.	It is not in the interest of top managers to tamper with the system that has promoted them. It is not necessary for top management to look at the big or whole picture.
16. We don't need constant informal parties as they do in Silicon Valley.	It is not necessary to foster an industry-wide culture of innovation, intense competition between companies, informal sharing of information, entrepreneurism, and the intense cycling of executives between firms.
17. We've discovered the principles of organization for all time.	Not only is the organizational structure of U.S. car makers appropriate for its time, if not all time, but it is well suited to responding to

(*continued*)

Table 5.1 (*Continued*)

Straight Talk	Professional Talk
	changing governmental policies, consumer tastes, and foreign competition.
18. No one, including ourselves, can teach us anything about good organization; we resist learning anything even from ourselves.	Despite GM's great success due to its early organizational structure under Alfred P. Sloan, Ford was correct to resist the professionalization of its upper management for so long, and Chrysler was correct to resist adopting GM's structure of high differentiation and high integration. In other words, U.S. car makers had nothing significant to learn from one another regarding the design of their respective organizational structures.

plain talk" interpretations of the more scholarly wording of the assumptions we have listed in column two.

As one looks down the list, one can see that the assumptions are so intertwined with technological, human, social, and organizational factors that it's literally impossible to say where one factor clearly leaves off and another begins. The entire list in effect constitutes a social contract for the automobile industry. Or to put it somewhat differently, the entire set of assumptions constitutes an orchestrated belief system. The individual assumptions neither exist nor make complete sense by themselves. They fit together and mutually reinforce one another as part of a larger pattern. Once again we see that the entities forming the basis of social reality are anything but "simple."

What's especially interesting is that in one way or another all the assumptions assume that nearly everyone connected with the "system" had no need for more than a fragmented, compartmentalized understanding of the business. *Ironically, although all the individual assumptions taken together constituted a "complex system," they had the effect of denying the very complexity of the system they were trying to manage.* It was also assumed that customer preferences were well understood and that customers really didn't want very sophisticated cars. All these may appear stupid to us now, but they weren't for a long time, almost sixty years.

The automobile industry failed to see that when their basic assumptions began to change, the industry needed to base its practices on new ones. The greatest difficulty, however, is that when a set of rules makes sense for so long, it's almost impossible to change those rules because they take on the character of eternal truths.

The moral is: The U.S. automobile industry did not fail because it was a failure from day one, but because it was a success for so long and it took its continued success for granted. The industry thought it had the magic formula for eternal success when actually all it had was a particular set of conditions good for only a limited period.

The failure of the U.S. auto industry is not that of a perpetual pattern of failure; instead, it is "the failure of success." Like most industries, instead of changing when it needed to—and in the best of all possible circumstances by anticipating the future—it did even more of what it did in the past. It reinforced and acted on its old assumptions and ways of doing things.

Now look at Table 5.2 which lists a particular set of assumptions pertaining specifically to GM. The first column outlines the generic set of issues that GM, like all organizations, had to manage. The middle column describes the old operating assumptions. In many ways, they merely repeat the assumptions contained in Table 5.1. We have listed them again so that we can contrast the old versus the new operating assumptions that we believe GM as well as every member of the automobile industry has to abide by.

The second and third columns of Table 5.2 constitute a dialectic. They represent two completely opposite, polarized views about the automobile industry. They differ fundamentally on nearly every one of the key dimensions that pertain to the meaning and running of a business. Further, although we believe the new operating assumptions make more sense for today's world, both views are plausible. Both in a way make sense, are coherent, internally consistent, and so forth.

Table 5.3 is a more systematic list of the basic issues that from our experience are involved in nearly all dialectics involving business. We have listed existential issues first. These are the deepest, core issues. In times of intense change, the key question, "What business are we in?" always comes to the surface.

The critical point is that *data have virtually no meaning on their own.* While this is true of all IS's, it only becomes readily apparent in the Dialectic IS. Thus, in this particular IS, the role of data takes on an entirely different status than in the Inductive-Consensual IS. In the Dialectic IS, the purpose of data is not to settle issues, but rather to surface the intense differences in background assumptions between two or more divergent positions. By seeing explicitly two positions operating on the same data set, we have the opportunity to witness systematically the background assumptions that the proponents of different positions

Table 5.2 General Motors: Assumptions and Counter-Assumptions

Generic Issues	Old Operating Assumptions	New Operating Assumptions
1. What business are we basically in? Who has basic control of the organization?	1. GM is in the business of making money, not cars. (The accounting and finance people took control of the organization after the industry passed its start-up phase, which was run by people who wanted to make cars.)	1. GM is primarily in the business of making quality cars, not money. Any organization that forgets its purpose for going into business in the first place will not achieve one of its fundamental financial objectives. (The engineers and the accounting/finance people should share control.)
2. What must our posture toward innovation be?	2. Success comes not from technological leadership but from having the resources to quickly adopt innovations successfully introduced by others.	2. We cannot give up technological leadership in a world that is more competitive than ever. We no longer have the luxury of time in a more complex environment.
3. How does the customer view our product?	3. Cars are primarily status symbols. Styling is therefore more important than quality to buyers who are, after all, going to trade up every other year.	3. Quality and styling are equally important in a more competitive market where even the cheapest car is expensive by past standards and where the competition is able to produce well-crafted and stylish products.
4. How much control do we actually have over our outside environment? How much can we really insulate ourselves from it?	4. The American car market is isolated from the rest of the world. Foreign competitors will never gain more than 15% of the domestic market.	4. The American car market will never be isolated from the rest of the world as it once was. Foreign competition is here to stay, and it will always be significant.
5. What are the basic resources this organi-	5. Energy will always be cheap and abundant.	5. Energy will never again be cheap or

(continued)

Table 5.2 (*Continued*)

Generic Issues	Old Operating Assumptions	New Operating Assumptions
zation needs to do business, and how available will they be in the future?		abundant, even though it may be held artificially low for what seems like an indefinite period of time; it will fluctuate enormously.
6. What skills and education of personnel do we need to presume?	6. Workers do not have an important impact on productivity or product quality.	6. Even with automation, worker attitudes and skills at all levels are more important than ever.
7. How isolated are we from the shifting concerns of our customers?	7. The consumer movement does not represent a significant portion of the American public.	7. Given the rising costs of all products and the increasing concern with the environment, there will continue to be some organizations that will represent these concerns. Any organization that ignores them is dangerously deluding itself.
8. What is our attitude toward the government? Who do we perceive to be our natural enemies, our allies; why?	8. The government is the enemy. It must be fought tooth and nail every inch of the way.	8. The government is a significant factor in the environment, and as such it must be dealt with whether one likes it or not. It is too easy to blame others for our problems.
9. Which types of controls are appropriate?	9. Strict, centralized financial controls are the secret to good administration.	9. Compulsive financial controls are the cause and effect of bad administration. There is all the difference in the world between a financial system that *controls* an organization and one that *enables* it to do what it wants to and should do.

(*continued*)

Table 5.2 (Continued)

Generic Issues	Old Operating Assumptions	New Operating Assumptions
10. How closed off is our organization to new ideas from the outside? How open, how trusting are we? What's our organizational culture like?	10. Managers should be developed from the inside.	10. The culture of an organization should be continually assessed to ensure that it has not become a closed system resistant to new ideas.

bring with them to convert data to information. To put it mildly, this is a very different use and interpretation of data or observations.

In the Inductive-Consensual IS, data, facts, or observations presumably have meaning and existence standing on their own. In the Multiple Realities and Dialectic IS's, the self-existence, self-standing, and isolated meaning of data is absurd. One has to use some model/theory/view of the world to gather data in the first place and, potentially, some

Table 5.3 Dialectic Issues

Generic Issues	Specific Issues

Existential
  Identity
  Meaning
  Purpose
  Ethical
	What business are we in?
	What is—ought to be—the purpose of our business?
	Who should we serve?

Stability
  Continuity
  Control
  Adaptability
	How stable is our core technology?
	How stable is the surrounding environment in which we operate?
	How much does our organizational structure need to change?
	What core information do we require?

Limits
  Scope of change
  Innovation
	How much change/innovation do we require?
	What ought to be the scope—limits—of our innovation? How extensive?

Stakeholders
  Consumers
  Managers
  Workers
  Government
  Special Interest Groups
  Suppliers
  Competition
	How stable/well known are consumer preferences?
	How much control do managers have/need to have?
	How much self-control do workers need to have? Morale?
	What is the relation between us and regulators? Adversarial/harmonious?
	Do our suppliers have a real stake in serving us?
	Friendly/fierce/adversarial/supportive?

other model/theory/view of the world to interpret it or make sense of it. In the meaning of "system" that we have been developing in this book, data are *not separable* from the underlying models, theories, or views of the world that are used to gather and interpret it.

We can begin to see that as in the case of the Multiple Realities IS, the inputs into a Dialectic are also complex. These inputs consist of the common data set plus the opposing assumptions (models) that characterize the deep positions of the two proponents.

Also, as in the case of the Multiple Realities IS, the decision-maker or observer of the debate is the operator in the system. He or she must adopt one of the two pure positions (sets of assumptions) or form a new position through synthesis or some other process as a result of witnessing the debate. This point is exceedingly important. Rarely, if ever, is the debate for the two initial contending positions. It is instead primarily for the decision-maker or observer who is not one of the initial adversaries. For instance, we doubt seriously that there is anything—almost by definition—that a "staunch" pro-life and pro-choice advocate could say to each other that would change their minds.

The Dialectic IS

In the Multiple Realities IS described in the last chapter, there was at least the possibility of some overlap between the range of models on an important issue. In the Dialectic IS, ideally, there is none. Further, the strongest debate occurs when both models—in essence, different sets of polarized assumptions on the key issues represented in Table 5.3—are restricted to the *same* data set in order to argue their respective positions. For instance, during the Vietnam War, a "hawk" and a "dove" could presumably both use the same data—number of Americans versus Viet-

If "information" is thus measured by the change in one's position—which it is in many scientific treatments of the concept of "information"—then little if any information passes back and forth between the two adversaries in a dialectical debate. This is why the staged debates between the two sides of the abortion controversy are no longer informative. We and they have heard it all before. On the other hand, if information is measured by the clarification and deepening entrenchment into one's own position, then the two poles do pass a kind of information, although it is not of the sort associated with the modification of one's position.

The guarantor of this system is conflict. It is hoped that as a result of witnessing an intense, explicit debate between two polar positions that the observer will be in a much stronger position to know the assumptions of the two adversaries and as a result clarify his or her own assumptions. It is also hoped that the observer or decision-maker will be in a stronger position to form his or her own position on a key issue. This of course presupposes the kind of observer or decision-maker who can tolerate and learn from conflict. In other words, there is no guarantee—as with all the other IS's—that the guarantor will actually work.

In this system, something is "objective" if and only if it is the product or the result of an intense debate between two polarized positions. (It is of course important to note that this is only this system's definition and preferred notion of "objectivity." This does not mean it will necessarily produce outcomes that will be "objective" from the vantage point of other IS's. Indeed, from their standpoint something will not necessarily be objective if it survives a debate.) Thus, the decision-maker or group of decision-makers might well form a single agreed upon position as a result of witnessing a dialectical debate. However, the agreement so reached is in principle very different from that of an Inductive-Consensual IS. It comes *after* an intense debate, not before or during.

Conclusions[3]

It is important to demonstrate that the Dialectic is not confined merely to technical or business problems—we can apply it to all the problems we face. Consider, for instance, homelessness. Suppose we reconceptualized homelessness as a problem that was amenable to business solutions, instead of a lost cause, or as only treatable by conventional, nonprofit public agencies?

We currently operate as if we think competition works only in the commercial marketplace and that the nonprofit world is one of altruism and good feeling. The rewards for participation in nonprofits are different, of course, but if one accepts the premise that "organizations are organizations" and those that do best are those that find a better way to pursue their tasks because they have a superior product, are more efficient at producing it, or have better defined a market for themselves.

Homelessness is a problem reaching epidemic proportions: many more people are now on the streets and they are a markedly different population from those who used to frequent skid rows across America. The old solutions, such as they were, no longer work.

Given this, what if we took a completely different approach and, in

---

[3]This section is based on an earlier paper by Ian Mitroff, Xandra Kayden, and Maxine Johnston, Director of the Weingart Center, Los Angeles, California.

the process, challenged some basic assumptions that have been endorsed for far too long? What if we thought of homelessness as a business problem instead of a lost cause that could only be treated by nonprofits?

A handful of new organizations around the country are doing exactly what successful businesses do—they are asking themselves what their clients want and are giving it to them. Instead of measuring the numbers of mouths they feed, they count the number of persons they can actually get off the streets. Many organizations claim to do this, but few actually deliver on it.

One such program exists in the center of Los Angeles's skid row—the homeless capital of the nation. It is called the Weingart Center and it functions out of a renovated twelve-story building, offering accommodations for 600 people. The resemblance between the Weingart Center and traditional programs ends there. Its walls are a cheerful yellow, freshly cut flowers grace the lobby table, a corporate flag waives proudly from the roof, and—most important—a complex of social service agencies, ranging alphabetically from Alcoholics Anonymous to Veterans Affairs, is on site, coordinated and managed in an entrepreneurial fashion.

The Center is not simply a place for the homeless to get a meal and spend the night; it is in the *business* of getting the homeless off the street and either back into society as self-supporting individuals or into long-term institutional care.

The Weingart Center did not drift into this business accidentally, any more than Federal Express drifted into the overnight package delivery business. Its approach came about as the result of painstaking planning and meticulous implementation on the part of the Center's current business-minded management, which took charge in 1985, when it was facing financial ruin.

Organizational crisis is often a precursor for significant innovation. Faced with bankruptcy, the new management concluded that in order to achieve its goal of addressing the problem of homelessness, it needed to create an integrated, multi-service, customer-driven organization built on several new assumptions:

- There is no one right or best way to serve the poor; different approaches would thus result in competition, which could serve the best interests of homeless people.
- The true measure of success is not the number of homeless people served or processed, but the number helped out of the *cycle* of homelessness.

- Homeless services ought to be regularized on a businesslike basis to encourage long-term institutional funding so that the organization can focus on helping its clients rather than on begging for money.
- Bringing people back into society requires a social setting—not only room and board but a cheerful environment and easy access to medical, mental health, job training, and placement services.
- The staff will take pride in an organization that is economically rewarding for themselves and serves their customers by giving them what they need. (The Center recently surveyed its residents to determine how they could be better served; many asked for wake-up calls so that they could get to medical appointments, job interviews, and other engagements on time.)

These assumptions are not typical of human service agencies, but are the foundation on which a remarkable success record has been built. More than 60% of the nonmentally ill people who have passed through the doors of the Weingart Center are permanently off the streets, living testimonials to the idea that even a social problem as difficult and as complex as homelessness is susceptible to innovative, entrepreneurial, and market-driven management approaches.

And this is the heart of the matter. The Center chose to operate entrepreneurially rather than bureaucratically. It decided to manage itself as a business rather than as a charity. It faced crises with imaginative leadership, including the adaptation of business management techniques such as a detailed business plan and emphasis on results as well as process, customer surveys, and an eagerness to be innovative. The Weingart Center is not the only, and perhaps not even the best, answer to homelessness. But it is a leader in a market long overdue for competition. Even more, it is ample testimony to what "new thinking" applied to old problems can accomplish.

In an age where increasingly all our social problems seem intractable, "new thinking" has become a catch-phrase. Unfortunately, most who propose it stop with the label. If the Weingart Center demonstrates anything, it is that new thinking is neither mysterious nor undoable.

PART III

NEW THINKING

CHAPTER
6

Unbounded Systems Thinking:
The Fifth Way of Knowing

We began our journey into the world of inquiry by introducing two familiar ways of knowing, *agreement* and *analysis*. Agreement relies on common observations typically gathered through the senses to define and structure problems. Analysis seeks to do the same by creating scientific models and studying their behavior. Both begin the process of solving problems with precise definitions of "the problem" and its boundaries, that is, its supposed separation and unique existence independent of other problems. Both believe that Truth exists and, further, that it is singular. Both seek simple, clear-cut solutions. Usually a single discipline, profession, or group of like-minded experts are used to give meaning and structure to a problem.

A Review of Analysis and Agreement

As the great Scottish philosopher David Hume, recognized, neither the method of Agreement nor that of Analysis can establish one of the most central concepts of science, causality. Thus, at the core of both Agreement and Analysis is a fundamental defect that cannot be corrected if we remain solely within the confines of these two systems.

To establish that one event, say A, is the cause of another, say B, we must first determine that A is both necessary *and* sufficient for the occurrence of B. First, whenever A alone is present, then B must occur. Whenever we flip light-switch A to the up position, then light B must always come on. If this always happens, then this establishes the sufficiency of A for B. Second, if whenever A is *not* flipped or on, then B must also not come on. This establishes the necessity of A for B.

Hume understood that mere observations alone (which are at the core of agreement) could never establish the logical necessity between two events. Observations of A and B as discrete *events* can never produce a *general concept*, that is, the logical connectedness between A and B; they cannot tell us what is *inherent* in the very *nature* of A and B that ties them inseparably together. For example, whenever we observe billiard ball A slam into billiard ball B with the resulting effect that A stops dead in its tracks and B moves away, we do not observe causality per se, which is a concept. We observe only the discrete events of A moving, A slamming into B, and B moving away. Or we observe that if A does not slam into B, then B stays put. To establish causality, we would have to observe in *every case* where A slammed into B, that A and A *alone* was sufficient for B to move, and, further, if A did not slam into B, then B would not move away.

Analysis fares no better. There is nothing in the *concepts* of A and B that necessarily connects them. We can think without logical contradiction of a set of properties for A, B, and A and B together, that does not bind them together causally. Thus, neither method of inquiry discussed in Chapters 2 and 3, agreement or analysis, can establish one of the most fundamental principles or conceptual relations in science—causality.

A Review of Multiple Realities

Next we introduced Multiple Realities. This way of thinking recognizes the limits of the single- and simple-minded processes of agreement and analysis. The Multiple Realities IS argued that one can never collect any data about any problem without having first presupposed some "model" of it, however implicit, vague, or intuitive. For instance, in Chapter 2, in giving their various estimates of steel production, the experts were relying on "something" to generate their numbers. These "somethings" may be as varied or as loose as different assumptions, different implicit models, or the extrapolation of past trends into the future. No matter what they used, the experts were not merely following implicit models or metaphors to *interpret* their estimates *after they already made them*. Instead, they were using, and indeed had to use, "something" to *generate* their numbers in the first place.

The Multiple Realities IS is convinced that multiple combinations of models *and* observations are more likely to lead to truth than any single model or set of observations. Consider, for example, the case of aerial photography. A single photograph may fail to grasp the nature of an object on the ground. But a series of photographs taken during a fly-over may yield a "better picture" of the object. As the shadows from different angles vary, it becomes clear that "the object" may be, for instance, an aircraft hangar. Similarly, two different models may combine to produce an explanation that neither can attain by itself. In physics, the behavior of light can only be adequately explained by calling on both particle and wave theories of light. Light may be modeled or described in the form of both particles *and* waves. Sometimes its particle properties provide more insight into its behavior; other times, its wave properties. Light is both a particle *and* a wave. The set of properties that predominate depend on the particular application or experiment.

As with every IS, Multiple Realities has serious problems. Multiple Realities needs "some initial sense" of what "the problem *is*" in order to produce an initial range of viewpoints to model "it." For instance, in the previous example, we have to have some initial idea that "the object" is a "building structure" so that we can position the camera correctly to seek photographs of "it" from different angles.

This difficulty is what Churchman has termed "Kant's problem"[1] to differentiate it from the much more commonly and widely recognized philosophical problem referred to as "Hume's problem," that is, the failure of either agreement or analysis to establish causality as one of the most central of all scientific concepts.

Kant showed that causality was not a property of things or objects themselves, but an integral feature of our minds. The concept of causality was somehow "built into" the structure of human minds (or must be presupposed) as a necessary precondition for humans to have experience. *Our minds must impose on the world, or presuppose, that which is necessary for experience to result.*

The very definition of "experience" means the *ability* as well as the *necessity* to perceive a "this" happening before a "that," and, among many other things, a "this" as the cause of a "that." Our minds are so structured that we are able to perceive "this's and that's," or objects and to recognize structure or connectedness between them. We are only able to find order in the outer world because there is a prior inner order or structure that is inherent in our minds.

Kant realized that for human beings to be able to observe objects existing in space and time, the mind had to have a "ruler" and a "clock" built into its mental apparatus as a necessary precondition. Since the very definition of experience means to perceive "something" existing "either here or there" and further at "this or that time," then a "ruler"

---

[1]C. West Churchman, *The Design of Inquiring Systems* (New York: Basic Books, 1971).

or a "clock" makes sense. In Kant's day, there was only one kind of "ruler" and only one kind of "clock" possible. Thus, there was no design problem regarding which one to choose. The "ruler" was that of Euclidean geometry; the "clock," Newton's physics or, more precisely, the concepts of space and time built into Newtonian physics. Today, however, we know that Euclid's geometry and Newton's physics are only very special cases of more general ones, for example, non-Euclidean geometry and Einstein's physics, to mention only two examples. A wide variety of different "rulers" and "clocks" exist that we can use to represent or structure events or problems in space and time depending on the problem. Thus, the question, "*Which type of representation* is relevant?" is very critical indeed. However, no one knows exactly how to specify fully the Executive, that is, the entity that selects the range and types of initial representations used to structure a particular problem. Hence, the existence of "Kant's problem," or "How do we choose a set of representations that are relevant for a particular problem?"[2]

Multiple Realities IS's also have some strong features in common. Multiple Realities IS's generally reflect multiple scientific and professional disciplines. Each discipline highlights different features of the problem at hand and in general produces very different observations or constructs very different models. Nonetheless, they share some characteristics:

- The models used are primarily technical in nature or have a strong technical bias.
- Every Multiple Realities IS, and certainly the technical models that generally compose them, presuppose the prior existence and primacy of the science of logic. The models in virtually all Multiple Realities IS's place an extremely high premium on rationality and logical consistency. All take for granted that an assertion or proposition cannot both be true and false at the same time.
- All of them see the world primarily in terms of "problems" and "solutions."
- They are concerned with the validation of precisely stated scientific hypotheses and the replicability of observations so that in principle "all qualified observers can observe or witness the same observations about a problem."
- All of them not only claim to be "objective" but place an extremely high premium on objectivity. Even though we demonstrated earlier that the concept of objectivity that is proper to this particular IS is the notion that objectivity results from the ability to witness a wide variety of ways

[2]Notice that these same questions also apply to analysis and agreement, yet they only become explicit with Kant. Thus, with the advent of the Multiple Realities IS, we not only become aware of the initial problems that plague this particular IS but all others as well.

of structuring a particular problem, most Multiple Realities IS's none-theless define objectivity in terms of the notion that "the properties of the observer shall not influence what he or she is describing." In other words, the observer is supposedly separable from what he or she observes. Of course, without an individual observer or modeler, there are no observations or models. Both are the products of human creativity. Especially in the social sciences, claims of objectivity are highly unrealistic, if not patently absurd: For instance, Churchman writes:

> One of the most absurd myths of the social sciences is the "objectivity" that is alleged to occur in the relation between the scientist-as-observer and the people he (or she) observes. He (or she) really thinks he (or she) can stand apart and objectively observe how people behave, what their attitudes are, how they think, how they decide. . . . [It is a] silly and empty claim that an observation is objective if it resides in the brain of an unbiased observer.[3]

As Mitroff has shown in a detailed study of the scientists who examined the rocks that were returned to earth from the Apollo moon missions, even physical scientists have serious doubts about the existence of "unbiased observers."[4] One of the most striking findings was that not only do scientists *not* believe in the existence of "unbiased observers," but they thought that the concept of an unbiased observer was actually harmful to the progress of science! Almost to a person, each of the forty-two scientists whom Mitroff interviewed repeatedly over the course of the Apollo moon missions argued strongly for the presence of sharply opposing views and scientists in order to advance the institution of science. In their more sophisticated view of science, the Apollo moon scientists believed that science progressed as the result of strong conflicting—even biased—claims being put forth. Objectivity, if it exists, is not a property of individual observers, for no human can ever be said to be purely objective or unbiased in his or her mental processes. Rather, if it exists, objectivity is a *systemic* property of the system of science taken collectively. Objectivity, if it results, occurs from a process of weeding out conflicting claims.

- In spite of their strong belief that single measures or single views of a problem are necessarily incomplete, Multiple Realities IS's still have the tendency to reduce all issues to numbers in the ultimate desire to produce a "single best" or optimal solution to a particular problem.
- They all practice reductionism, simplifying an apparently complex system by subdividing it into subsystems and repeating this process until they arrive at a problem they can solve. In the words of von Foerster's First Law:

---

[3]C. West Churchman, *Challenge to Reason* (New York: McGraw-Hill, 1968).
[4]Ian I. Mitroff. *The Subjective Science: A Philosophical Inquiry into the Psychology of the Apollo Moon Scientists* (Amsterdam: Elsevier, 1974; reissued by Intersystems Publishers, Seaside, CA, 1984).

The more complex the problem that is being ignored, the greater are the chances for fame and success.[5]

"Success" is defined mainly in academic terms, that is, as a plethora of papers that deal with the solution of unimportant and even trivial problems. Other techniques of simplification include the linearization of a nonlinear problem and the use of averages to hide extreme cases.

* Since Multiple Realities IS's primarily represent problems in technical terms, issues of aesthetics and ethics are largely separable, nonexistent, or even meaningless, since in general these cannot be reduced to numbers or technical models.

A Review of Conflict

Next, we consider another IS, *Conflict*. Here "truth" evolves from the confrontation of opposites, between a thesis and an antithesis. The most familiar example of conflict in action is the American courtroom. A prosecutor presents the testimony of witnesses to construct the strongest case possible for proving the guilt of a defendant. The defense attorney on the other hand then presents witnesses to build the strongest case possible for the innocence of the defendant. The task of the jury is to reach a verdict, presumably the "truth of the case." The word "truth" is perhaps not entirely appropriate here since the law is more concerned with "justice" than with truth.

An executive may use a similar dialectical process when faced with an important decision. He or she may ask one group to make the strongest case in favor of entering a new line of business; another, to make the strongest case against it. He or she then reaches a decision by having the two groups challenge their respective assumptions and argue their positions before the executive. More often than not, professionals with non-science or non-engineering backgrounds are able to use the Dialectic IS more effectively than those with science or engineering backgrounds.

It should be apparent by now that Agreement, Analysis, Multiple Realities, and Conflict as IS's all have strict limits. We should also realize that the first three are more closely associated with the sciences and technical professions, while the Dialectic is used much more readily in others.

Unbounded Systems Thinking (UST)

We now want to focus on a fifth IS, which in principle sweeps in all the others. We label it *Unbounded Systems Thinking* (UST) and contend that

---

[5]H. von Foerster, "Responsibilities of Competence," *Journal of Cybernetics*, Vol. 2, 1972, pp. 1–6.

it is the basis for the "new thinking" called for in the Information Age. This does not mean that this particular IS is perfect in the sense that it is free from significant defects or problems. Indeed, we can be confident that no IS devised by humans will ever be free from all defects.

We will find that in UST "everything interacts with everything," that all branches of inquiry depend fundamentally on one another, and that the widest possible array of disciplines, professions, and branches of knowledge—capturing distinctly different paradigms of thought—must be consciously brought to bear on our problems.

In UST, the traditional hierarchical ordering of the sciences and the professions—as well as the pejorative bifurcation of the sciences into "hard" versus "soft"—is replaced by a circular concept or relationship between them. *In UST, every one of the sciences and professions is considered fundamental; none is superior to or better than any other.* (For a more detailed explanation, see the Appendix.) The basis for choosing a particular way of modeling or representing a problem is not governed merely by considerations of conventional logic and rationality. It may also involve considerations of justice and fairness as perceived by various social groups and by considerations of personal ethics or morality as perceived by distinct persons.

In UST, we find that the supposed distinct and separate existence of the various IS's that was implied in the preceding chapters is a fiction. Consider Analysis, for example. The models one creates are strongly influenced by the background and experience of the particular modeler, particularly his or her training and previous modeling experiences. Thus, an analyst trained in economics will favor econometric models and focus on econometric variables. Environmental considerations are usually dismissed as "externalities," that is, outside the model. Further, all values tend to be measured in economic units such as dollars. The original "world dynamics" models unconsciously reflect Western culture as well as the engineering background of the modelers. Thus, considerations not contained purely in the models nonetheless have a real and significant bearing on them.

For another, consider the Delphi Method described in Chapter 2. The numerical judgments of a panel of technical experts with regard to a particular problem reflect strongly the scientific and technical models they are familiar with. Consequently, they unconsciously endorse the same assumptions that underlie accepted models and arguments. In this way, Analysis subtly insinuates itself into Agreement.

Consider another possibility, the charge that Agreement is quick to level at Analysis. Agreement points outs that at Analysis' core is an unreflected or unconscious Inductive-Consensual IS. All the basic "self-evident" propositions that Analysis starts with are only "self-evident" to a very particular group of *one*, that is, the person doing the Analysis. The basis for all of Analysis's inquiries is not a set of pure propositions uncontaminated by experience, as Analysis believes, but Analysis's unconscious experience or judgments.

In a similar manner, it can be shown that *every IS or inquiry presupposes all other ISs.* At some point in its operation, every IS presupposes some fundamental concept or process from each of the other IS's. *In this sense, all IS's are interdependent or mutually dependent on one another.* (For a fuller explanation of this point, see the Appendix.)

Because UST is the most difficult of all the IS's to grasp, we're going to lead up to it in two stages. First we'll give a very brief and general overview of the systems approach. Next, we'll use a concrete problem-solving method, or vehicle, that shows how to apply it.

The Systems Approach: General Background and Overview

In 1896, the great American philosopher William James of Harvard University wrote a letter to Provost Harrison of the University of Pennsylvania recommending Edgar Arthur Singer for a position in philosophy at his institution. James wrote that in his thirty years of teaching philosophy, Singer was the "best all around student" that he had had "in the philosophic business." There was no aspect of philosophy that Singer could not do well.

Singer went on to a long and distinguished career in American philosophy. Among his many outstanding students was C. West Churchman from whom Mitroff studied philosophy of science at the University of California, Berkeley, in the 1960s.

The point of this all too brief bit of history is not just that the first author can trace his intellectual lineage back to one of the world's most distinguished philosophers, and one whom both authors admire greatly, but that Singer was one of the most important participants in the founding of the modern systems approach. Churchman in turn extended Singer's ideas significantly and their ideas form the philosophical basis for the modern systems approach.

It would take us too far afield to discuss Singer's and Churchman's philosophy in detail,[6] but we can give a good sense of it by means of a seemingly simple example.

Before he became a philosopher, Singer was a civil engineer. (And no one is more practical than civil engineers, as the first author knows all too well since he too was one early in his career.) In Singer's day (the

---

[6]See C. West Churchman, *The Design of Inquiring Systems* (New York: Basic Books, 1971).

1890s to the 1950s) most scientists and philosophers thought that the act of measuring the distance between two points on the surface of the earth, say *A* and *B*, was among the simplest of all the operations in science. Presumably, all one did was to stretch out a ruler or tape measure on the ground and merely read the correct result. Given the immense complexity of modern physics and engineering in the twentieth century, surely this simple act of measurement paled in comparison to the great feats of modern science.

First, the measurements had to be read by a human observer. And, if anything is characteristic of humans, it is that they make mistakes. This means that the " 'simple' act of reading a tape" is not as simple as it first appears. We have to know or "sweep in" psychological knowledge about how and why humans perceive the ways that they do and what kinds of errors they typically make so that we may correct for them. Otherwise, we will repeat what the great astronomer Bessel did in the nineteenth century.

Bessel had several research assistants who worked for him. Their task was to time the passage of certain stars, or the movement of the heavens, between two hairlines in a telescope. When the position of certain stars crossed the first hairline, an assistant started a clock and then stopped it when it crossed the second. Naturally, the times differed significantly from one research assistant to another, and Bessel wanted to fire the slowest and the fastest. If the science of psychology had been as well developed in his time as it is now, and if Bessel were inclined to use it, then he wouldn't need to dismiss anyone. If he had known about the concept of "reaction times"—the fact that people differ systematically in the times it takes them to react to a red or a green light—then he would have been able to correct their differing times or "standardize" the measurements of each observer by applying a correction factor. That is, he would have been able to add a correction factor to the measurements of those who were too slow and subtract another factor from the times of those too fast.

The simple act of measuring the distance between two points *A* and *B*, which supposedly is only an exercise in physics or engineering, in actuality encompasses or presupposes knowledge from the science of psychology as well. Are the observers working under stress? If so, then this will influence what they see and read. Are they tired, cold, hungry, upset, happy, or depressed? Do their jobs or an important contract hinge on their measurements? Are human lives and safety at stake so that the measurement becomes extremely critical and important to verify? What is the purpose of their measurements? All these issues and more are embedded in this seemingly simple act.

In a similar fashion, Singer showed that *every one* of the sciences and professions known to humankind was involved in the act of measuring the distances between *A* and *B*. For instance, if the distance was long enough, then it had to be broken up and measured in stages by two or more different observers. But this means that there had to be cooperation between them and even an overall management plan for accomplishing the task. This in turn meant that social sciences such as sociology and even political science were now involved. Further, since the task had to be managed, professions such as management were also involved.

It is well known that group phenomena are in general very different from individual. Groups can influence their individual members to take positions that by themselves none of the individuals may support. That is, groups can take positions that are both riskier or more conservative than any of their individual members. How does one member of a group who is involved in the act of measurement influence others, both consciously and unconsciously? This question is not trivial. Groups can systematically distort both the gathering and the resultant interpretation of important numbers, especially if significant outcomes hang in the balance.

Suppose that the measurement of the distance between *A* and *B* is extremely critical, say for political reasons such as redistricting. It would not be unreasonable to employ different firms or teams of surveyors to measure the distance repeatedly and see if the same "answer" keeps emerging. But this only complicates matters. In general, there are distinct, systematic differences between different scientific laboratories or companies due to inherent differences in their corporate cultures or general ways of thinking. Thus, different laboratories tend to come up with different results even when they are all using standardized procedures! Thus, once again, we need to include other sciences or professions to help account for the differences between measurements.

The idea that the measurement of the distance between two points is *solely* a matter for the physical scientist or engineer would be amusing if it were not so out of touch with social reality. We challenge anyone to tell two neighbors who are fighting over a disputed property boundary line that the problem is only "scientific." The problem is as likely to be historical as well as legal and political.

The upshot of Singer's analysis was that *there were no elementary or simple acts in any science or profession to which supposedly more complex situations could be reduced.* Every act or action performed by humans was complex and therefore had within it a complex series of other actions. Furthermore, unlike the scientists and the philosophers of his day who believed that some sciences such as mathematics or physics were the

most basic or fundamental, Singer believed that there were no fundamental sciences to which all others could be reduced. Since it was
necessary at some point to involve every science in the actions of every
other science, all the sciences and professions were equally fundamental. No single science stood at the top of the totem pole or hierarchy of
science and in essence, every science depended on every other.

This fundamental notion of interconnectedness, or nonseparability,
forms the basis of what has come to be known as the Systems Approach.
In essence, the Systems Approach postulates that since every problem
humans face is complicated, they must be perceived as such, that is,
their complexity must be recognized, if they are to be *managed properly*.
Notice the emphasis on the critical words "managed properly." As a
critical human activity, science, or the creation of a very special kind of
knowledge, must be conceived of and *managed as a whole system*.

Mitroff served as a policy consultant to the U.S. Bureau of the
Census[7] on another problem also assumed to be one of the simple
"acts" in science. This consultancy showed that the counting of something so large as a population as huge, as diverse, and as dispersed as
that of the United States is another complex systems problem. In particular, it demonstrated that the act of counting is only a "simple act" if one
sticks to simpleminded examples for which little consequences hang in
the balance. Thus, it may or may not be a critical task to count all the
pencils on one's desk depending on the circumstances and the uses to
which the number will be put.

Most people conceive of counting as a simple act because they have
placed it in a context that is either trivial or for which no important
consequences result. Thus, in a completely sealed off room of small
dimensions, it may well indeed be a trivial task to count the number of
marbles or playing blocks in front of a small child. Counting is thus seen
as simple because most people have in their heads an example drawn
from childhood where everything of relevance was directly before them
or under their control. Such examples in no way represent the difficulty
of counting because all the true complexity, and hence the real problem
itself, has been eliminated by design.

A population as large as the United States's, approximately 250
million, is as much a theoretical construct as it is a "real thing." First, no
one ever has or ever will "see" all 250 million people at once. Second,
what is indeed *the real population* at any particular time? Each instant,
"some" are in the "process" of being born while others are in the

---

[7]See Ian I. Mitroff, Richard O. Mason, and Vincent P. Barabba, *The 1980 Census: Policy Making amid Turbulence* (Lexington, MA: Lexington Books, 1983).

"process" of dying. Third, many people, such as illegal aliens, the poor, the homeless, the transient, the politically disaffected, criminals, etc., don't want to be counted at all. Since a complete count of all persons is mandated by the U.S. Constitution, how can we count or estimate these individuals?

Fourth, it is known that the census misses blacks and Hispanics more than whites.[8] On the average for the United States as a whole, the census undercounts whites by 2% and blacks by 8%. Should the number of whites therefore be multiplied by 1.02 and blacks by 1.08 to correct the "raw numbers"? The answer is "No!," for the numbers 2% and 8% only reflect national averages. In some areas of the country, blacks are missed less and whites by more than the national averages. Thus, one needs to know what the percentages are at local levels. But to know this requires money and time, since counting is not cost-free as it is for those who assume or conceive of it in simpleminded terms.

In 1980 the counting of the U.S. population cost roughly $1 billion. Since the census is the basis for the distribution of over $100 billion from the federal government to the cities and the states, as well as congressional apportionment, it is a very serious exercise indeed. Whenever large amounts of money and votes are at stake, this gets the attention of politicians and voters real fast. Thus, in 1980 over fifty separate localities sued the U.S. Bureau of the Census to correct or adjust the "raw" census counts in order to better reflect the "true" numbers of persons in their districts so that they could receive a larger share of congressional apportionment and revenues due them. The suits were largely unsuccessful. There was, however, a humorous part to them that reveals another systemic aspect of counting.

Since it was cheaper and more effective to bring one suit instead of fifty separate ones, the fifty localities and states banded together in a single class-action suit. At the end of the suit, the court costs naturally had to be divvied up. Guess which numbers were used. There was no other basis than the original raw census counts of the populations of the localities that were in dispute. Thus, a set of numbers that were not good enough for one purpose—congressional appointment and revenue sharing—were good enough for another—dividing up court costs. And this indeed is the whole point—counting depends on one's purposes. Whether a set of numbers is appropriate depends on the purposes for which those numbers will be used. For some purposes a rough estimate is sufficient; for others, we need to be as "precise as possible within cost constraints." Thus, no single set of numbers will satisfy all purposes.

This example also shows clearly that in a complex system there is no

[8]*Ibid.*

way of getting completely outside it to reach a purely objective number that will satisfy all parties.

The design and conduct of a census, and the interpretation and use of census numbers, are obviously not simple. Counting the population of the United States is not therefore an exercise in statistics alone. Although statistics do play a central role, equally important issues of ethics, the law, and politics enter as well. Thus, once again as in Singer's supposedly simple problem of measuring the distance between two points A and B, every science and profession enters into the conduct and interpretation of census numbers.

With this as background, we're ready to turn to a description of Unbounded Systems Thinking.

UST and The Multiple Perspective Method

We will describe UST in terms of a somewhat new problem-solving method known as the Multiple Perspective Concept or Method. Let us begin with the differences between Multiple Realities or *Model/Data Couplings* and *Multiple Perspectives*. A perspective is typically much broader than a model/data coupling. A particular set of model/data couplings is generally bound by the background philosophical values of science: the reduction or breakdown of all complex problems into a set of smaller ones; the assumption that for every problem there exists a solution; the belief that everything important can be reduced to numbers, and so on. In contrast, Multiple Perspectives represents Unbounded Systems Thinking. It includes other, distinctly different, paradigms or ways of thinking. In the Multiple Perspective Concept, what we have previously called Agreement and Analysis are collectively subsumed under the label of the *Technical or T Perspective*. In addition, two other major types or kinds of perspectives also exist.

Consider three examples.

J. Forrester has applied techniques from electrical engineering, specifically systems dynamics, to model a corporation. His approach is known as Industrial Dynamics. It provides important insights about corporations, especially the flows and materials within various subunits. In contrast, the brilliant Florentine civil servant Niccolo Machiavelli in *Discourses* (1531) and *The Prince* (1532) offered very different insights concerning organizations and especially their control. Machiavelli's insights have been translated not only from the Italian into English, but from an ancient idiom into a modern one in Anthony Jay's book, *Management and Machiavelli*.[9]

[9]Anthony Jay, *Management and Machiavelli* (New York: Holt, Rinehart & Winston, 1948).

Both Machiavelli and Forrester are looking at organizations, but from very different angles. Similarly, Gareth Morgan views organizations in terms of various "metaphors." Thus, organizations can be viewed as "machines," "living systems," "cultures," and "political systems." Each metaphor provides a valuable way of thinking about complex systems.[10]

Using three very different perspectives, Graham Allison examined the Cuban missile crisis during JFK's administration: (1) the rational actor perspective, (2) the organizational process perspective, and (3) the bureaucratic politics perspective.[11] In the rational actor perspective, an analyst considers the "United States" and the "Soviet Union" as single decision-makers. Each pursues national goals, considers various alternatives for action, and chooses those particular alternatives that from a rational standpoint "maximize" its objectives. The organizational process perspective recognizes, on the other hand, that each government is not monolithic, but composed instead of various suborganizations, each with its own parochial priorities and perceptions. The bureaucratic politics perspective sees governmental actors not as whole organizations, but rather as individual players defined by their positions and personal interests.

A corporate executive must make an important decision on a new line of business. There is a detailed cost-benefit analysis from the executive's technical staff recommending a single, "optimal" alternative or solution. The executive does not make the final decision solely on the basis of the report. The executive talks to various department heads to determine whether there is strong support or opposition for the proposal. Over the weekend, the executive also talks with an old friend who runs a company in a different field. Valuing that person's judgment, the executive discusses his or her ideas. In addition, the executive draws upon intuition and experience. Only then does our executive decide in his or her own mind—presumably without a formal or mathematical weighting formula—between several different, probably conflicting, perspectives: technical, organizational, and personal or individual.

Each of these examples has a crucial characteristic in common: *Each perspective reveals insights about a problem that are not obtainable in principle from the others.* In particular, the organizational and personal perspectives draw on or *sweep in* societal and human aspects that pervade all complex systems.

The Multiple Perspective Concept involves three very different types of perspectives in addressing complex problems.[12] However, we must

[10]Gareth Morgan, *Images of Organizations* (Beverly Hills, CA: Sage Publications, 1986).

[11]Graham Allison, *Essence of Decision: Explaining the Cuban Missile Crisis* (Boston: Little Brown, 1971).

[12]Harold A. Linstone, *Multiple Perspectives for Decision Making* (New York: Elsevier, 1984).

emphasize that we are merely augmenting, not replacing, the T perspective. Specifically, we employ:

> T: The Technical Perspective.
> O: The Organizational or Societal Perspective.
> P: The Personal or Individual Perspective.

O and P bring to the forefront human beings collectively and individually in all their complexity. From the O perspective, the group or organization considered may be a formal or informal organization and range in size from that of a family to a global network, from a local government to a transnational corporation, from a small research team to a world religion. The difference in perspectives forces us to distinguish *how* we are looking from *what* we are looking at. Each incorporates distinct sets of underlying assumptions and values.

Table 6.1 summarizes some of the main features of each of these distinct world views or perspectives. One of the strongest implications of the table, if not that of the whole Multiple Perspective Concept itself, is that its use is not a luxury. All complex problems—especially social ones—involve a multiplicity of actors, various scientific/technical disciplines, and various organizations and diverse individuals. In principle, each sees a problem differently and thus generates a distinct perspective on it.

Inherent in the use of the Multiple Perspective Concept are the following key characteristics:

- The systems designer, analyst, or manager is a fundamental part of the system or problem being analyzed; that individual's psychology as well as social context are inseparable from how he or she represents a system or a problem. (See the Appendix for a more detailed discussion.)
- The choice of which particular perspectives to bring to bear in a specific problem or to emphasize is a matter of one's ethical values and judgments, even though all complex problems invariably involve all three perspectives. In essence, we have another version of "Kant's problem," that is, "How do we choose a set of *perspectives* that are appropriate for a particular problem?" In the earlier example of the executive seeking various inputs for his or her decision making, how does that executive justify the choice of analyses, departments, and personal friends consulted? Individual ethics and value judgments are implicit in the selection process as are one's scientific and professional background, training, and so on. Thus, strong considerations of ethics, values, judgments, and background experiences underlie the Multiple Perspective Concept, and, indeed, all applications of UST.

Table 6.1 The Three Multiple Perspective Types and Their Paradigms

	Technical (T)	Organizational (O)	Personal (P)
World view	Science-technology	Social entity, small to large informal to formal	Individuation, the self
Goal	Problem solving, product	Action, stability, process	Power, influence, prestige
Mode of inquiry	Sense-data, modeling, analysis	Consensual and adversary	Intuition, learning, experience
Ethical basis	Logic, rationality	Abstract concepts of justice, fairness	Individual values/morality
Planning horizon	Far	Intermediate	Short, with exceptions
Other characteristics	Looks for cause and effect relationship	Agenda (problem of the moment)	Challenge and response
	Problem simplified, idealized	Problem delegated and factored	Hierarchy of individual needs
	Need for validation, replicability	Political sensitivity, loyalties	Filter out inconsistent images
	Claim of objectivity	Reasonableness	Need for beliefs
	Optimization (seek best solution)	Satisficing (first acceptable solution)	Cope only with a few alternatives
	Quantification	Incremental change	Fear of change
	Trade-offs	Standard operating procedures	Leaders and followers
	Use of averages, probabilities	Compromise and bargaining	Creativity and vision by the few
	Uncertainties noted (on one hand . . .)	Make use of uncertaintes	Need for certainty
Communication	Technical report, briefing	Language differs for insiders, public	Personality important

- The value in using multiple T, O, and P perspectives lies in their ability to yield unique insights. None by itself suffices to deal with a complex system, but together they give a richer base for decision and action. Using a single perspective may be compared to employing a single dimension to depict a three-dimensional object. Each added dimension facilitates comprehension. On the other hand, the ability to live comfortably with and tolerate multiple perspectives is extremely difficult for many people. No matter how much additional insight they provide, some people simply cannot tolerate three very different perspectives of a single problem.
- Any complex problem may be viewed from any perspective. For example, an organizational decision may be seen from a T perspective; technology may be viewed from a P perspective (see the Appendix). Furthermore, two organizations concerned with the same problem may have quite different O perspectives on it.
- O and P differ in fundamental, key characteristics from T. As a result, O and P inexorably move us beyond those features associated with basic science and engineering. At best, the scientific method operates within the framework of the T perspective. However, science cannot prove that any particular model offers the "absolutely correct" representation of reality; it cannot give assurance that the variables chosen are sufficiently inclusive or appropriate.
- In the Multiple Perspective Concept, we also cannot prove that a set of perspectives is the "right" set any more than an executive can prove that he or she involved the "right sources" before making an important decision. We cannot derive the "correct weighting" in integrating perspectives any more than a jury can in integrating the testimony of different witnesses to reach a decision.
- Two perspectives may reinforce one another or cancel each other; frequently, they interact in a dialectic mode.
- A perspective may change over time. The T perspective changes when a fundamental scientific revolution occurs, such as when the Copernican model of the solar system replaced the Ptolemaic one. In the P perspective, a persuasive individual may impose his or her individual perspective on an entire organization. This most naturally occurs when a strong leader is at the helm. The personal perspectives of a Napoleon, a Lincoln, an FDR, a Hitler, and a Stalin, as well as those of Henry Ford, Thomas J. Watson, Jr., Edwin Land, Soichiro Honda, Frank Lorenzo, and Lee Iacocca became in effect the O perspectives of their respective organizations.
- It is not easy at times to distinguish between an O and a P perspective: is the person giving his or her own or the organization's perspective? Effective organizations are usually characterized by a strong congruence between the O perspective and their members' P perspectives.
- In "real-life" situations, managing problems consists of at least three activities: (a) analyzing alternatives, (b) making decisions about which

alternative to choose, and (c) successfully implementing the chosen alternative. The T perspective focuses most strongly on (a) and least on (c); hence, the "gap" so often deplored between analysis and action. Successful implementation depends first and foremost on the use of human resources and this means that O and P become crucial as we move from (a) to (c).

The world as seen from a pure O perspective is ideally an orderly progression from state to state, with occasional minor crises along the way, for which experience and procedural manuals have "all the answers." Rules and "standard operating procedures" are drawn up and followed. Once these are promulgated, policy becomes quasi-sacred. The O perspective also reflects the culture and the myths that have helped to mold and bind the organization, group, or society together as a distinct entity in the eyes of its members. All companies have a strong corporate culture. However, cultures are especially strong in those organizations that are particularly outstanding or poor. Human beings have a great need for illusions and O perspectives often provide them. They give a sense of belonging to a superior chosen group and a readiness in its members to sacrifice for the common good.

The importance of understanding the difference between O perspectives has been pointed out by M. Thompson with reference to the nuclear power industry.[13] Scientists and Engineers for Secure Energy (SE) was a pro-nuclear *caste;* the Sierra Club, an anti-nuclear *caste;* and Friends of the Earth (FE), an anti-nuclear *sect.* There are inherent and important differences between castes and sects. Castes are hierarchical, whereas sects are egalitarian. Castes deal with multiple issues; sects, with a single issue. Castes are stable when mature; sects, inherently unstable. Castes are national or global in scope; sects, local. Castes tend to use science in conventional ways; sects use "facts" only when they support their pet issues or projects. Castes are willing to compromise and negotiate; sects are not. Knowledge of these different perspectives are helpful in developing strategies for dealing with castes and sects. For example, if SE wants to diffuse FE, it should recognize that it can negotiate with other castes, such as the Sierra Club, but not with FE. By seeking out anti-nuclear castes, it can begin a dialogue. It must also correct distortion caused by media coverage of sects. Communications technology has vastly altered the power of a sect—the media can magnify a sect's impact to an incredible degree. A minuscule sect can alert the media of a demonstration, and the small size and short duration are completely masked in the typical two-minute television evening news

---

[13]M. Thompson, "Fission and Fusion in Nuclear Society," *Newsletter of The Royal Anthropological Institute,* Vol. 41, 1980.

segment shown to millions. The resulting amplification of sect influence can create paralysis in the political establishment. Grass-root computer networks may well become the sects of the future.

The understanding of O perspectives is obviously central to the question of reducing the gap between technological and societal change. They can not only help to develop practical policies (e.g., identifying possible new coalitions), but can also focus attention on the tremendous role that social processes play in the physical sciences, such as the allocation of research resources to prestigious institutions and individuals (see the Appendix). They can draw out impacts not apparent from the use of other perspectives, such as new realities that are constantly being created within organizations.

P perspectives are equally significant. They are often the most subtle, the most elusive, and the most difficult to specify. Here the world is seen through the eyes and the brains of particular individuals. Intuition, charisma, leadership, and self-interest play vital roles in matters of policy and decision and can only be understood through the P perspective. From Adam Smith to West Churchman, concern has been expressed over the danger of ignoring the individual and losing him or her in the aggregate. Over 200 years ago, Adam Smith wrote:

> The man of system . . . seems to imagine that he can arrange the different members of a great society with as much ease as the hand arranges the different pieces upon a chessboard; he does not consider that the pieces upon the chessboard have no other principles of motion besides that which the hand impresses upon them; but that, on the great chessboard of human society, every single piece has a principle of motion of its own altogether different from that which the legislature might choose to impress upon it.[14]

Churchman writes:

> Economic models have to aggregate a number of things, and one of the things they aggregate is you! In great globs you are aggregated into statistical classes. . . . Jung says that, until you have gone through the process of individuation . . . you will not be able to face . . . social problems. You will not be able to build your models and tell the world what to do. . . .
>
> From the perspective of the unique individual, it is not counting up how many people [are] on this side and how many on that side. All the global systems go out: there are no trade-offs in this world, in this immense world of the inner self. . . . All our concepts that work so well in the global world do not work in the inner world. . . . We have great

[14]H. W. Schneider (Ed.), *Adam Smith's Moral and Political Philosophy* (New York: Hafner Publishing Company, 1948).

trouble describing it very well in scientific language, but it is there and it's important. . . .

To be able to see the world globally, which you are going to have to be able to do, and to see it as a world of unique individuals . . . that is really complexity.[15]

Freud perceived three layers of social life: the professional, the political, and the personal. He found the first to be the most current and readily accessible, the third the deepest, least current, and least accessible. The analogy with our three perspectives should be clear.

A human being's own personal P perspective is of course itself a product of genetic and environmental influence. There is no formula, no assurance of balance. We describe some people as "analytical," others as "organization men," and still others as "strong personalities." In other words, one perspective often seems to dominate a person's unique mix and thus characterizes him or her uniquely.

An example of irrationality is the strong tendency of individuals to neglect both feedback and complexity. Human beings simplify situations by focusing on one-dimensional, one-way effects. They ignore two-way linkages between events and feedback loops. Military planners are especially prone to this tendency in studying future enemy strategies and tactics; they develop responses to enemy strategies but then fail to consider what the enemy might do in return to minimize the effectiveness of the initial strategies. Technologists generally start with a problem and then develop a solution. However, they neglect to take the next step and probe for the problems created by their solutions. *They fail to analyze how T solutions to T problems become the O problems of the next go around, and vice versa.* The P perspective recognizes the strong presence of such behavior in complex systems.

Intuition is as well-appreciated trait in the business world as the following quotes suggest.

On each decision . . . mathematical analysis only got me [a CEO] to the point where my intuition had to take over.

Walk through an office, and intuition tells you if things are going well.[16]

Effective CEOs are aware that rationality and the scientific method provides critical inputs to only one of three critical questions that overarch key decisions. These are: (a) does it add up? (b) does it sound OK? and (c)

---

[15]C. West Churchman, "A Philosophy for Planning," in *Futures Research: New Directions,* H. A. Linstone and W. H. C. Simmonds (Eds.) (Reading, MA: Addison-Wesley Publishing Company, 1977).

[16]Quoted in R. Rowan, "Those Business Hunches Are More than Blind Faith," *Fortune,* April 23, 1979, p. 112.

does it feel right? Logic and science contribute primarily to the first question, less to the second, and even less to the third.[17]

Highly respected scientists also have not been afraid to acknowledge the role of intuition and accept it as an important method. In his classic, *The Psychology of Invention of the Mathematical Field,* Jacques Hadamard wrote:

> That those sudden enlightenments which can be called inspirations cannot be produced by chance alone is already evident. . . . There can be no doubt of the necessary intervention of some previous mental process unknown to the inventor, in other words, an unconscious one. . . . It is quite natural to speak of a more intuitive mind if the zone where ideas are combined is deeper, and of a logical one if that zone is rather superficial.[18]

These words echo Freud. Most recently, Nobel Laureate Herbert Simon and his associates have explored the differences between experts and novices in solving physics problems. They found that the expert is mentally guided by a large number of patterns serving as an index to relevant parts of a "knowledge store." The patterns are:

> Rich schemata that can guide a problem's interpretation and solution and add crucial pieces of information. This capacity to use pattern-indexed schemata is probably a large part of what we call physical intuition.[19]

Each individual has a unique set of patterns that informs his or her intuition. In drawing on the P perspective, we are thus augmenting conscious, logical T processes by opening ourselves to the deeper mental levels that store patterns of potential value. But this also means that there are important limitations to so-called "expert systems" that are being designed by "knowledge engineers." Their goal is to translate completely into rules how experts solve problems so that the rules can be programmed into computers. This desire to reduce complex knowledge that, by definition, encompasses T, O, and P into strictly T terms may work well for problems that are bounded and structured. However, it cannot be used for more complex systems and problems where an expert must of necessity call on patterns that are largely unconscious.

The P perspective has an important characteristic that differentiates

---

[17]R. G. H. Siu, "Management and The Art of Chinese Baseball," *Sloan Management Review,* Spring 1978, p. 85.

[18]Jacques Hadamard, *The Psychology of Invention of the Mathematical Field* (Princeton, NJ: Princeton University Press, 1945), pp. 21, 113.

[19]J. Larkin, et al., "Expert and Novice Performance in Solving Physics Problems," *Science,* Vol. 208 (4450), June 20, 1980, p. 1342.

it strongly from T or O. It often serves as a far better channel of communication. Novelists and playwrights are well aware of this as are politicians. George Bernard Shaw used Eliza Doolittle and her father to portray class problems in England. In *War and Peace*, Tolstoy felt it essential to present the massive invasion of Russia through the eyes of individuals, not organizations or abstract forces. He recognized that a focus on Napoleon would be totally inadequate, that each individual mattered. His unresolved dilemma was his inability to integrate the P perspectives of the vast number of participants. He could not do what, say, the physicist does in dealing with the immense number of particles in a gas, that is, integrate the individual effects statistically into an over-arching general physical law. Even so, Tolstoy's use of the P perspective was highly effective. Similarly, the enormity of the Nazi Holocaust cannot be properly conveyed by the T perspective that uses statistics such as the over six million people murdered to depict its horror and evil. Instead, the enormity of the Nazi Holocaust is better conveyed through a P perspective such as *The Diary of Anne Frank* or the TV production, *Holocaust.*

Former President Reagan understood well the power of highly personal anecdotes. In the business world, Peters and Waterman also state:

> We are more influenced by stories (vignettes that are whole and make sense in themselves) than by data (which are, by definition, utterly abstract).[20]

The multiplicity of perspectives is vital, but, as noted earlier, it introduces ethical concerns, and, for a great many, discomfort as well. How does one integrate perspectives? How does a decision-maker finally combine all the inputs he or she obtained to make a decision?

Typically, many business executives find it impossible to specify their decision processes in detail and how they go about integrating various sources.

How does a jury arrive at a decision? The prosecutor has presented witnesses and integrated their perspectives to deliver a summation to the jury. So presumably has the defense attorney using a different combination of witnesses. The jury may accept either summation or go back to the testimony itself and attempt to do its own integration. There is no neat scientific or T methodology for accomplishing this task.

We have pointed to two inherent concerns in using the Multiple Perspective Concept: the selection of initial perspectives and their integration. All executives wrestle with such questions daily—often success-

---

[20]T. J. Peters and R. H. Waterman, Jr., *In Search of Excellence* (New York: Harper and Row, 1982), p. 61.

fully—even though they often do it intuitively. It can be argued that the selection of the proper perspectives constitutes *the* test of effective decision making and implementation. While many such individuals do not necessarily need advice in applying multiple perspectives, others can benefit from a few guidelines:

1. *Strive for a balance among T, O, and P perspectives.*   Individuals with strong bias toward analysis, such as engineers and scientists, are more likely to spend an excessive, if not obsessive, proportion of their time with T, with which they are comfortable, and treat O and P as superficial addenda, if indeed they are even so gracious as to consider them as addenda at all.

2. *Use "good" judgment in selecting perspectives.*   There are as many O perspectives as there are "organizations within an organization." Within a company or agency, each separate department or division has, in principle, its own unique O perspective. One cannot therefore include them all. In addition, many O perspectives are often in conflict. Thus, the analyst is well advised to construct a Dialectic between different internal O perspectives in order to draw out different insights into the complexities within an organization. These same cautions apply to P perspectives. The hierarchy of an organization is not always a good guide to follow; key persons do not always appear on formal organization charts. Individuals who act outside of an institution may be as important and can affect its outcomes greatly.

3. *In obtaining information, recognize that O and P require greatly different methods than T.*   The T perspective generally relies on the analysis of scientific data and models. One-on-one interviews have proven advantageous in revealing what makes an organization and individual actors "tick." Structured questionnaires or Delphis are no substitute for direct interviews. Talented interviewers must almost by definition be good listeners, sensitive to nuances and nonverbal communication. What is not said is often as important as what is. Volunteered asides may be as significant as direct responses to formal questions. Academics with their generally pronounced T bias often make poor interviewers in soliciting O and P views. O and P become especially critical in understanding other cultures.

   For instance, Linstone's research in China revealed that the Chinese generally understood well what was being probed for with O-type questions. Chinese culture is bureaucratic and hierarchical so that O games and strategies are generally known to everyone. Power relationships are enshrined in all kinds of slogans such as "two down, one up" for the process of planning and "the pyramid of power." P perspectives present more of a hurdle. Often, the answers to such questions are extremely sparse so that one must push for concrete examples and anecdotes both to flush out and flesh out critical insights. All translation becomes interpretive and thus requires sophisticated knowledge of the local culture. Since simple word-for-word translations are not possible, and Chinese culture contains many untranslated metaphors, similies, and allusions, extremely well-trained and sophisticated interpreters are absolutely essential.

4. *Pay particular attention to the mutual impact, interdependencies, and integration of perspectives.* This is one of the most critical steps in the entire process. We cannot reiterate enough that we are dealing with Unbounded Systems Thinking. There is no formula, no pat procedure to guarantee or to assure that all interactions are taken into account or that their perspectives are weighted correctly.

5. *Beware of thinking statically in dynamic environments.* In its ideal form, UST is non-terminating; it recognizes the dynamic and ongoing nature of the real world. Actors, both organizational and individual, come and go, and change their perspectives over time. Even traditional operations analysts, such as RAND's Charles Hitch, have recognized that every significant problem they tackle had to be redefined once their analysis was underway. The decision process involving the location of liquified energy gas facilities in several countries has been examined.[21] In each case, the decision process was sequential and could be divided into several discrete rounds. The actors as well as the various definitions of the problem changed from one round to the next. An understanding of the sequential nature of the process proved critical in developing insights into the problem.

   Real-life sequential decision processes can also lead to weaknesses. By changing actors and definitions of the problem or problems across rounds, important feedback and opportunities for cross-cueing may be missed. For instance, a multiple perspective study of the 1980 California Medfly Eradication Program found this to be a serious handicap in the decision process.[22]

Conclusions Regarding the UST IS

Of all the ISs we have encountered thus far, UST is the most difficult to represent. Its various "components" are the most complex and are more highly linked than in any of the ISs we have examined thus far. Recall that in the Multiple Realities IS, at least the possibility of "some overlap" existed between the various models of an important problem. In UST, there is no longer the "mere possibility" of "some overlap" between different models, but strongest of all, *every model of a problem we*

[21]H. C. Kunreuther, et al., *Risk Analysis and Decision Processes* (Berlin: Springer Verlag, 1983).
[22]H. Lorraine, "The California 1980 Medfly Eradication Program: An Analysis of Decision Making Under Non-Routine Conditions," *Journal of Technological Forecasting and Social Change*, Vol. 40, 1991, pp. 1–32.

pick to represent it presupposes in an important sense every other model we could have used to represent it. Thus, *in UST every element or component is strongly inseparable from every other element or component.* This is expressed in Churchman's definition of a "problem":

> Something is a problem if and only if it is a member of the set of all other problems.

In other words, something is a problem if and only if within it is contained every other problem. Or, conversely, something is a problem if and only if its definition and solution both lead to and follow from the definition and solution of every other problem. To put it mildly, the "inputs" into UST are "messy" indeed.

UST generalizes the sweeping-in process that we have referred to throughout this chapter and which is discussed in more detail in the Appendix. The Multiple Perspective Concept uses the notion of "sweeping in," or adding two other critical perspectives, the O and the P, to the natural tendency in Western cultures to reduce all problems and issues to T issues. UST consists of an even more radical sweeping-in process.

UST extends this to its most extreme or radical limits—in principle, UST sweeps in *every* discipline, profession, way of knowing so as to give the broadest possible view of any problem. Just when we think we have zeroed in on the solution or the definition to a critical problem, UST picks another perspective to open up the whole problem even further.

This sweeping-in process that is a fundamental characteristic of UST also differentiates it from other IS's. For example, at their best, Multiple Realities IS's are inter- or multidisciplinary systems. They seek to integrate the knowledge from different professions or scientific disciplines into a coherent, overall view of a problem, without necessarily changing the structure of those basic disciplines or professions. UST, on the other hand, is much more radical in its approach; it is fundamentally a *trans*disciplinary IS. That is, UST does not believe that we can "solve" important problems by respecting the current structure of the disciplines, professions, or the modern university. The various sciences and professions are the product of human organization. In this sense, they are artifacts, not natural entities. As such, in working on every problem, we are also simultaneously working on how we organize ourselves to solve our problems.

Because all problems are invariably linked with one another in complex ways, we can take this as an excuse for hopelessness and helplessness. However, we can also view it as an opportunity, as a sign for hope. The fact that all problems invariably involve one another does not mean as the simpleminded critics of UST often imply that "one must

know everything before one can know anything." Instead, the unbound-
edness of all problems, of all systems, can be construed as an opportunity and a challenge to perpetually enrich our knowledge of the world.
Not every IS is compatible with the personality of every problem-solver.
How one views UST is thus in part dependent on the individual. People
differ radically, one of the very points of the Multiple Perspective Concept. Some thus regard UST as a rich resource; others, as something to
be avoided at all costs. We believe that as natural as both of these
reactions are, the way of knowing we have outlined in this chapter,
palatable or not, is increasingly important.

Finally, the Multiple Perspective Concept has been applied by
Linstone and associates to a wide spectrum of "messy" systems ranging
from a corporate acquisition and integration of a new technology into a
business unit, to the prediction process in a public power administration,
a perinatal health care program, regional development in the Third
World, the decision to drop the atomic bomb, the M-16 rifle development, technology impact assessments, and industrial risk management.
In each case the multiple perspectives sweep in crucial insights that
could not have been obtained by reliance on the T perspective. The next
chapter discusses an important case of the management of a risky technology in a foreign culture, the Bhopal chemical castastrophe.

PART IV

APPLICATIONS AND METHODS OF NEW THINKING

CHAPTER
7

Bhopal: Catastrophe Making*

Humankind has always faced hazards. Prior to the twentieth century, natural disasters, such as earthquakes, constituted the major danger. Today, modern hazards are increasingly caused by humans themselves—fires, explosions, chemical spills, drugs, pollution, radioactive material, and so on. What happened in Bhopal, India, is richly illustrative of the dangers posed by modern technologies and for this reason, a multiperspective analysis of it is important if we are to learn how to avoid future disasters.

Bhopal: General Background

In 1984, the world chemical industry was highly competitive. Although Union Carbide was an important company in the industry (37th largest in the United States), employing almost 100,000 employees with plants in forty countries, it had strong pressures for increasing profitability and reducing costs. That year the firm realized an after-tax profit of only about 60% of its major competitors. Also, as part of a major strategic shift, Union Carbide divested about forty strategic business units in the late 1970s and was considering selling its 50.9% controlling interest in Union Carbide (India) Ltd. (UCIL).

Like its parent company, UCIL faced several difficulties. Although it

*The authors would like to acknowledge the valuable contribution of Dr. B. Bowonder of the Administrative Staff College, Hyderabad, India, to this work. Portions of this chapter were originally published in *Technological Forecasting and Social Change*, Vol. 32 (1987), pp. 183–202, and are reproduced by permission of the publisher. Copyright 1987 by Elsevier Science Publishing Co., Inc.

was the twenty-first largest company in India, and employed more than 10,000 people at the time of the accident, the plant operated at less than 50% of total capacity. Faced with a highly competitive Indian pesticide market, itself in decline, and the strong pressures of its parent company, UCIL tried to take advantage of economies of scale. For example, it developed production of MIC, a highly toxic gas used in the fabrication of pesticides, although the company lacked industrial expertise in this area.

At the cultural level, the plant encountered great turnaround of top management who were demotivated in part by the potential divestiture of UCIL. In addition, many new managers hired lacked expertise in the chemical industry. Further still, UCIL cut personnel, lowering even more the morale of the company and compromising its security and emergency procedures.

For a long time, the Indian government had encouraged the large-scale use and production of pesticides as part of its efforts in the "green revolution" and the development of a stronger national industrial basis. The government was thus reluctant to place a heavy safety burden on industries, fearing a decrease in additional job opportunities. The government had also encouraged industrial development of Bhopal and its surrounding area.

Literally overnight, Bhopal was transformed from a feudal and traditional community in 1959 to a large industrial city. Attracted by jobs, Bhopal's population grew from about 100,000 people in 1961 to 670,000 in 1981, a rate 300% higher than the average growth rate in India. The exponential population increase resulted in a severe lack of housing facilities and a declining city infrastructure such as poor water supply, inadequate transportation, poor communication, poor education, health facilities, and so on. For example, at the time of the accident, Bhopal had only 37 public telephones, 1,800 hospital beds, and 300 doctors for a population of 670,000. This severe housing shortage was also influential in the development of a number of concentrated slums in the area, with thousands of people literally living across the street from the UCIL plant.

In this (simplified) context on December 2, 1984, the Bhopal catastrophe occurred. The initiating or precipitating *incident* was a leak of a highly toxic gas—MIC—that originated in the UCIL MIC unit. This incident was itself contextually related with several different factors and escalated into a major crisis. For the sake of simplification, we list here only five of the most important.

First, neither the Indian government nor UCIL's top management or employees were knowledgeable about the potential dangers of MCI production. For example, prior to the incident, an inadequate storage of MIC and other hazardous materials was allowed to multiply, an emergency plan was nonexistent, and employees lacked proper training and resources for facing an emergency. Also, the Indian government had only 15 inspectors for the 8,000 plants located in the province and the two inspectors assigned to UCIL had no training in chemical engineering. *Second,* many technical malfunctions and human errors were confounded. For example, several safety valves in the MIC unit were removed or did not operate during the incident; an emergency refrigeration unit was under repair and thus not available for use; an audio emergency signal to warn the outside population was turned off. *Third,* the immediate population of Bhopal was ignorant that they were in any danger; the general impression was that the plant was involved in the production of "plant medicine"; thousands of people living directly across the street from the plant received the full impact of the toxic gas. *Fourth,* the local authorities, also ignorant of the nature of the danger, commanded the population to flee from the city, not knowing that a better strategy would have been to direct people to lie on the ground and breathe through a damp cloth, thus alleviating the effects of the gas. Similarly, the health community, ill prepared for such an emergency, treated only symptoms, lacking adequate equipment and resources to treat the full health consequences that developed. *Fifth,* and last, the poor infrastructure of the city contributed in general to a relative lack of effectiveness of emergency efforts.

All these factors taken together created a total context that contributed to the death of between 1,800 to 10,000 people, and the injury of between 200,000 to 300,000, depending on which sources of information we use. In addition, incalculable effects were produced on animals and the environment. It is impossible to pinpoint specifically what "caused" the accident as well as what "caused" the death and the injuries of so many victims. Certainly, the lack of commitment and interest of Union Carbide to UCIL stands out; the general lack of knowledge and preparation of UCIL's management and employees; the technical failures and human errors; the push by the Indian government to develop Bhopal as an industrial area with its resulting lack of infrastructure and safety procedures; the fact that thousands of people were living directly across from the plant and were ignorant of its potential danger; the lack of knowledge and preparation on the part of local authorities as well as health specialists.

Risk from a T, O, and P Perspective

Table 7.1 depicts how the T, O, and P perspectives define and approach the analysis of risk. This table shows that disagreements about the nature of acceptable risks are deeper than simpleminded explanations

Table 7.1 Risk Concerns Seen in Perspective

Technical (T)	Organizational (O)	Personal (P)
One definition of risk for all	Definition customized to organization or group	Individualized
Compartmentalizing problem by discipline	Compartmentalizing problem by organization slot	Ability to cope with only a few alternatives
Data and model focus	Perpetuation of entity is the foremost goal	Time for consequences to materialize (discounting long-term effects)
Probabilistic analysis; expected value calculations	Compatability with standard operating procedures (SOP)	Perceived horrors (cancer, AIDS, Hiroshima)
Statistical inference	Avoidance of blame; spread responsibility	Personal experience
Actuarial analysis	Inertia; warnings ignored	Influenced by media coverage of risk (The China Syndrome)
Fault trees	Fear exposure by media; attempt stonewalling	Peer esteem (drugs)
Margin of safety design; fail-safe principle	Financial consequences	Economic cost (job loss)
Quantitative life valuations, cost-benefit	Impact on organization power	Freedom to take voluntary risks
Validation and replicability of analysis	Threat to product line	Salvation; excommunication
Failure to grasp "normal accidents"	Litigious societal ethic	Influence of culture
Intolerance of "nonscientific" risk views	Reliance on experts, precedent	Ingrained views; filter out conflicting input
Claim of objectivity in risk analysis	Suppression of uncertainties	Opportunity to gain respect; fame

such as "miscommunication" or "misinformation." Organizations and individuals "see" hazards from very different perspectives. As as result, they focus on very different risks. Organizations with different standard operating procedures formulate different ways of dealing with the same hazard. For instance, after a 1979 report indicated that formaldehyde causes cancer in rats, the CPSC banned its use, whereas the EPA and OSHA took no action.[1]

[1]N. Ashford, W. Ryan, and C. Caldart, "Law and Science Policy in Federal Regulation of Formaldehyde," *Science* 222:894–900 (November 25, 1983).

Why Go Beyond the T Perspective?

A major problem with the T perspective is that conventional probabilities become irrelevant when we deal with the conjunction of a large number of (a) very low likelihood events and (b) very severe consequences. The expected value (the probability of an event multiplied by its consequences) of the cost of an event as the product of two such quantities loses meaning. Consider, for example, the causal factors involved in Bhopal. Over thirty separate factors can be enumerated as potential causes. In addition, there was the time of occurrence (Sunday night) and the wind direction (south). The joint probability of the occurrence of all these causal factors at the same time is infinitesimal. Multiplication of such a probability of occurrence by, say, 10,000 fatalities, thus results in a very small expected value. In Perrow's terms,[2] this constitutes a "normal accident." It is an accident that by this logic is not severe. Engineers do not usually focus on such types (see Table 7.1). Thus, nuclear reactor engineers have a "preoccupation with large-break accidents . . . and an attitude that if they could be controlled, we need not worry about the analysis of 'less important' accidents."[3]

Not surprisingly, one reason for failure in design is:

• The nonexistence of a meaningful calculus, that is, a basis for calculation, for low-likelihood, high-consequence events.

A major characteristic of complex systems is that, in general, they involve many more uncertainties than technologists admit. Technical analyses rely primarily on data and models. These data are sometimes contradictory, often determined more by their availability or that they are variables that can be measured rather than their significance. Models simplify—and oversimplify—the real world. It can be shown that the number of possible relations among subsets in a system comprising only three elements is at least 49. In general, one has $(2^n - 1)^2$ relations, where n is the number of elements. A system comprising ten elements thus may involve at least 1 million possible internal relationships—and even this is not an upper limit. For example, the count assumes that there are no multiple interactions between the same subsets. Most connections are ignored in the hope that they do not exist or are insignificant.

[2]C. E. Perrow, *Normal Accidents* (New York: Basic Books, 1984).

[3]J. Kemeny, et al., *Report of the President's Commission on the Accident at Three Mile Island* (New York: Pergamon Press, 1979), p. 9.

As a result, another important design problem or issue arises:

- Our inability to consider the enormous number of possibilities inherent in complex sociotechnical systems.

In addition, an even more subtle design issue exists:

- We fail to recognize that complex natural systems—most obviously *homo sapiens*—are *safe-fail*, not fail-safe.[4]

Complex natural systems are designed to minimize the cost of failure rather than its likelihood. Ecological systems sacrifice efficiency for resilience; they trade the avoidance of failure for the ability to survive and recover from failures. Herein lies an especially crucial implication for human-made systems involving very low probability plus very severe consequence combinations. They cannot be made fail-safe because of the many "normal accident" possibilities inherent in them; therefore, it becomes desirable to make them safe-fail. A major question then becomes: What do we do if they cannot be made safe-fail? In other words, what if it is impossible to be certain that no intolerably severe accident can occur?

Ostberg[5] has suggested other related reasons for the failure to design for very low likelihood or "inconceivable" events:

- Our tendency to equate the very low likelihood of occurrence of an event with the impossibility of its occurrence—human beings do not and cannot deal well with quantities that are either very small or very large; the subatomic and cosmic both lie outside the scale of human action; we can distinguish mathematically or conceptually between probabilities of 10^{-2} and 10^{-5}, but we cannot do so psychologically.
- Inconceivable events are normally considered "soft"; as a result, verbal or qualitative signals typical of soft factors are not perceived by scientists and engineers accustomed to quantitative or "hard" signals.
- The lack of willingness and the means to learn from experience.

The complex and close intertwining of human and technical factors in most systems reveals the shortcomings of analyses based solely on the T perspective.

Many of the causal factors in Bhopal involved "human error." These include the use of cast iron pipes instead of specified stainless steel

---

[4]C. S. Holling, "The Curious Behavior of Complex Systems: Lessons from Ecology." In H. Linstone and W. H. C. Simmonds (Eds.), *Futures Research: New Directions* (Reading, MA: Addison-Wesley Publishing Company, 1977).

[5]G. Ostberg, "Evaluation of a Design for Inconceivable Event Occurrence." *Materials and Design* 5:88–93 (1984) and G. Ostberg, *On Imaging Surprises* (mimeograph) (Lund, Sweden: University of Lund, July, 1986).

pipes; removal of refrigerant from the MIC tank despite instructions to the contrary in the safety manuals; the failure to recognize the entry of water in the MIC tank; the failure to sound adequate warnings to the surrounding area for over three hours after the leak began. There is a striking similarity in this respect to Three Mile Island and Chernobyl.

At Three Mile Island, errors included inadequate training of utility company operators and supervisors, toleration of poor control room practices, failure of the construction engineers (Babcock and Wilcox) to inform their customers of persistent breakdown of pilot-operated relief valves. The President's Commission concluded:

> . . . the fundamental problems are people-related problems and not equipment problems . . . wherever we looked, we found problems with the human beings who operate the plant, with the management that runs the key organization, and with the agency that is charged with assuring the safety of nuclear power plants.[6]

At Chernobyl, a mishap on April 26, 1986, occurred in the context of a turbine test. Faulty actions caused a loss of the water that continuously cools the uranium fuel rods in the reactor's core. This led to a partial core meltdown. As of August 1986, thirty-one fatalities were noted and over 200 were hospitalized with radiation sickness. A total of 135,000 people had to be evacuated. The long-term effect is estimated at 5,000 to 24,000 additional cancer deaths in the Soviet Union over the next 70 years.[7] Although Westerners have pointed to the technical flaws in the reactor design,[8] the Soviet report to the International Atomic Energy Agency focused on a series of human errors, mistakes that violated safety regulations and, in some cases, common sense. Andronik Petrosyants, head of the State Committee for the Use of Atomic Energy, said:

> For almost 12 hours the reactor was functioning with the emergency system switched off. . . . It is quite possible that the [previous] smooth operations brought on complacency and that this led to irresponsibility, negligence, lack of discipline and caused grave consequences.[9]

Valeri Legasov, first deputy director of the principal Soviet atomic research institute, added:

> If at least one violation of the six would be removed, the accident would not have happened. The engineers psychologically did not believe that

[6]J. Kemeny, et al. *Report of the President's Commission on the Accident at Three Mile Island* (New York: Pergamon Press, 1979), p. 8.

[7]*Washington Post*, August 30, 1986.

[8]*New York Times*, August 26, 1986.

[9]*Washington Post*, August 22, 1986.

such a sequence of improper actions would be committed. Such a sequence of human actions was so unlikely that the engineer did not include [it] in the project. Is that human or technical?[10]

The six errors were:

1. The most crucial one: the operational reactivity margin at the reactor core was dropped substantially below the permissible level (6–8 rods equivalent instead of 30); result: ineffectiveness of the emergency protection system.
2. The power of the reactor was allowed to drop below the 700-megawatt level prescribed for the tests; result: the reactor was difficult to control.
3. During the experiment, all eight main circulation pumps were switched on, with individual pump discharges exceeding specified limits; result: lowered water levels, coolant temperature in main circulation circuit approaching saturation temperature.
4. Automatic blocking devices were shut off in an attempt to prevent the reactor from shutting down (so the testing could be continued); result: no automatic reactor shutdown capability.
5. Defense system controlling water level and steam pressure were blocked off; result: reactor protection systems cut off.
6. The emergency cooling system was shut off over a 24-hour period in violation of regulations; result: inability to reduce the scale of the accident.[11]

In both Three Mile Island and Chernobyl, the engineers had focused on the T perspective to the neglect of the O and P perspectives. The design of both plants had specifically considered technical failures, resulting in large-scale consequences, but underestimated the significance of nontechnical failures precipitating a chain of events inducing compound technical failures with large-scale consequences.

Proposals to correct problems that result from previous technical solutions nearly always involve more "technological fixes." They usually add to the complexity of the system and, hence, create new sources of failure. The situation is reminiscent of the inability to eliminate computer programming errors.

Consider a large computer program. At time zero, when it is completed, the program displays a errors. By correcting them, new errors are introduced, resulting in a maximum of b. After much work, a value c is reached. As the number of instructions increases above 10^4, c rises exponentially. This means that a deterministic but very complex algorithm has a behavior that is not completely predictable. And no gener-

[10]*Washington Post*, August 22, 1986.
[11]*Nuclear Engineering*, October, 1986, 31/387, p. 3.

al program exists that can correct the errors of any program. *The more complex the system, the more unpredictable it becomes.*[12] This consequence applies to hardware and software systems. In the case of a chemical plant, numerous monitoring systems would need to be stacked on a multitude of process control units; monitoring systems would have to oversee other monitoring systems. The complexity of the whole and the sources of failure increase as hierarchies are inevitably added.

A 1984 National Academy of Sciences study[13] found that adequate information exists on potential health hazards for only 18% of the 1,815 pharmaceuticals studied, 10% of 3,350 pesticide ingredients, and 11% of other commercial chemicals that were considered. These figures make the uncertainties surrounding MIC seem quite "normal." Some of the following questions could not be answered immediately after the accident:

- How could the effects of exposure to MIC be mitigated?
- Could vegetables and other food products exposed to MIC be consumed without detoxification?
- Was the water in the Bhopal lake near the UCIL plant[14] safe for drinking?
- What kind of health monitoring system should be instituted for monitoring MIC exposure effects?

Another aspect of uncertainty is confusion under pressure. In military terms we have "the fog of battle." Communications break down; human beings operating at great stress make unwarranted assumptions. The crisis in Europe preceding World War I is a classic example.[15] In looking at today's sophisticated strategic command and control system, the State Department's George Ball reminds us:

> The decisions of politicians and ultimately military commanders are never—and will never be—made in a sterile environment or dictated solely by mathematical possibilities. They will reflect the probability of over-hasty, poorly calculated responses, the pressure of alarmed and uninformed public opinion inflamed by propaganda and factual error, the fear, ambitions, frustrations, and anger of military and political leaders playing by quite different rules, acting and reacting on the basis of rumor and misinformation.[16]

[12]J. C. Simon, "Complexity Concepts and the Limitations of Computable Models." *Technological Forecasting and Social Change* 13:1–11 (1979).

[13]*New York Times*, December 16, 1984.

[14]B. Bowonder, "The Bhopal Accident," *Technological Forecasting and Social Change*, 32(2):169–182 (1987), Figure 1.

[15]B. Tuchman, *The Guns of August* (New York: Macmillian, 1962).

[16]G. W. Ball, Book Review, *New York Review of Books*, p. 5 (November 8, 1984).

At Three Mile Island, communications between the Nuclear Regulatory Commission and Three Mile Island management were less than satisfactory from the beginning. Harold Collins, the assistant director for emergency preparedness in the Office of State Programs, testified:

> I think there was uncertainty in the operations center as to precisely what was going on at the facility and the question was being raised in the minds of many as to whether or not those people up there would do the right thing at the right time.[17]

The operators in the plant's control room were unaware for hours that a critical valve was stuck open and draining cooling water out of the reactor. Joseph Hendrie, the Chairman of the Nuclear Regulatory Commission, and Governor Richard Thornburg of Pennsylvania were unsure whether to evacuate the area.

> His information is ambiguous, mine is nonexistent and—I don't know, it's like a couple of blind men staggering around making decisions.[18]

An O Perspective: The Corporation

A number of considerations are swept in by the O perspective with regard to Bhopal.

Proprietary Information

Much of the information on MIC was considered proprietary by UCIL. In particular, there was inadequate dissemination of data regarding its toxicity. According to the medical journal *Lancet,* nothing appeared in the medical literature during the seven years prior to the accident that would have provided insights on the effects of large-scale exposure of humans to MIC. U.S. Occupational Health Guidelines[19] for MIC indicated that MIC vapor is an intense lacrimator; it irritates the eyes, mucous membranes, and skin. Exposure to high concentrations can cause cough, dyspnea, and chest pain. Another study reports that MIC can cause permanent eye damage and is dangerous when inhaled even in greatly diluted form.[20] UCIL provided little indication of MIC toxicity

---

[17]J. Kemeny, et al., *Report of the President's Commission on the Accident at Three Mile Island* (New York: Pergamon Press, 1979), p. 118.

[18]J. Hendrie, quoted in D. Ford, *The Button: The Pentagon's Strategic Command and Control System* (New York: Simon and Schuster, 1985), p. 88.

[19]NIOSH/OSHA, *Occupational Health Guidelines for Chemical Hazards,* Report No. 81-123 (Washington, D.C.: Department of Health and Human Services, 1981).

[20]ACGIH. Documentation of the American Conference of Governmental Industrial Hygienists (Pittsburgh, 1984).

when it applied for a license for production of carbaryl pesticide and for expansion of its Bhopal facilities in 1976. The state director of public health knew nothing about the MIC handled at the plant; neither did the mayor, who was a physician, nor the chief administrative officer responsible for Bhopal disaster contingency plans.

One result of the lack of detailed clinical information was a medical controversy after the accident about the efficacy of using sodium thiosulfate as an antidote. The Head of Toxicology of Bhopal Medical College advocated sodium thiosulfate injections[21] while the Madhya Pradesh Department of Public Health opposed the treatment. Some toxicologists[22] thought that MIC decomposes to hydrogen cyanide at temperatures above 200° Centigrade and were concerned about combined exposure to MIC and hydrogen cyanide.

Stonewalling

The first organizational reaction to a technological disaster is almost always the same. The O perspective sees an industrial disaster as a threat to the organization. Therefore, there is a desire to suppress information and effect a coverup. H. J. Heinz stonewalled when the rancid tuna crisis erupted; Metropolitan Edison Company stonewalled when Three Mile Island occurred; the Soviet government stonewalled Chernobyl. It is almost a reflex action, a standard operating procedure (SOP), to circle the wagons when a threat looms. This reaction itself creates serious undesirable consequences, if not further crises. In the case of Chernobyl, it caused health concerns, if not panic, in Sweden, Poland, Romania, and Austria.

In Bhopal, the possibility of cyanide poisoning was first denied, alarms were delayed, and thousands received unnecessary exposure to the MIC cloud. In its "investigative report" Union Carbide Corporation asserted that 1,000 to 2,000 pounds of water would have been required to account for the chemistry of the residue left in the tank after the accident. It still insisted that the source of the water was unknown.[23]

Contingency Planning

It was known that MIC reacts violently with water and that water was a commonly used liquid in the plant. The possibility of an unintentional mixing was unlikely—it was not, however, impossible. Here is an exam-

[21]Rameseshan, R. Bhopal Gas Tragedy: Callousness Abounding. *Economic and Political Weekly*, 20(2):56–57 (1985).

[22]Morehouse, W. and Subramanian, A. *The Bhopal Tragedy.* New York: Council on International and Public Affairs, 1986.

[23]Union Carbide Corporation, *Bhopal Methyl Isocyanate Incident: Investigation Team Report* (Danbury, CT: Union Carbide Corp., March 1985).

ple of the low likelihood plus severe consequence combination discussed earlier. Traditional corporate planning tends to ignore remote possibilities. In their reliance on the T perspective, corporate analysts accept econometric models without probing core assumptions and adopt technological risk models that shun "normal accidents." In both areas uncertainty is anathema. It is either dismissed as being unmanageable or replaced by probability-type risk.

However, some corporations have learned a lesson from the military; they perform enough "what-if" exercises to develop the capability to handle surprises. Not only do they examine several alternative future environments but they also develop hedging strategies.[24] Beyond this, they are learning the art of "crisis management."[25] H. J. Heinz, after its rancid tuna shipping incident, has set up its own emergency-management team. Such firms no longer wait for a disaster to strike; they develop detailed contingency plans to cope with surprises. United Airlines has a crisis team prepared to take charge in the event of a major airline disaster. Dow Chemical has produced a 20-page program to communicate with the public during a disaster; it even includes such particulars as who runs the copy machines. There are consulting firms to train corporate officials and provide expert crisis managers. Existence of a trained crisis team at UCIL might have limited the extent of the disaster, thus resulting in fewer lives lost.

Management Inertia

A safety audit in 1982 at UCIL's Bhopal plant had pointed to the failure to use protective slip blinds before washing pipe lines. Nevertheless, workers continued this practice without corrective action. Early warning signals of potential crises are commonly ignored. One analysis of hazard warning structures has found that major hazards usually involve a series of small accidents prior to the occurrence of a major catastrophe.[26] Between 1978 and 1983 there were three toxic spills at UCIL's Bhopal plant. An inquiry into the 1981 accident was undertaken by the State Government, but no action was taken on the recommendations. A series of early warning signals as well as in-house safety audits conducted in 1979 and 1981 did not move UCIL management to take adequate action.[27]

[24]W. Ascher and W. Overholt, *Strategic Planning and Forecasting* (New York: John Wiley and Sons, 1983).

[25]"Coping with Catastrophe," *Time*, p. 53 (February 24, 1986).

[26]F. P. Lees, "The Hazard Warning Structure of Major Hazards," *Transactions of the Institute of Chemical Engineers*, 60:211–221 (1982).

[27]B. Bowonder, "The Bhopal Incident," *The Environmentalist*, 5:89–103 (1985); B. Bowonder, et al., "Avoiding Future Bhopals," *Environment*, 27(7):6–16 (1985); and A. Subramanian and J. Gaya, "Towards Corporate Responsibility," *Business India*, 202:42–54 (1985).

A UCIL worker who monitored the tank temperature comments:

For a very long time we have not watched the temperature. There was no column to record it in the log books.[28]

Companies use public relations to calm (or lull) citizens. U.S. utility industry reactions to Chernobyl included statements that such an accident cannot happen here because "defense in depth" design makes "our nuclear plants among the safest in the world."[29] Such statements are true but irrelevant: "safest in the world" may be unacceptable if the consequence of an accident occurrence is seen as catastrophic.

Some transnational companies maintain different standards for U.S. and overseas facilities or operations. Often this is done with the government's approval. In November 1986, President Reagan allowed the export of drugs that have not been approved by the Federal Drug Administration for U.S. sale.[30]

Blind Technology Transfer

Technology is not transferable in isolation—it takes place from one societal/cultural setting to another. In the case of the United States and India, vast differences exist between these settings and they affect the success of the transfer. The T perspective fails to capture such differences.

Human societies have unique cultural characteristics. American technology transferred to Japan develops in a unique way: Japanese-designed cars are readily distinguishable from American models. Japanese manufacturing operations differ from American ones. American engineers designing a system for American use rarely consider foreign applications in their work. Usually it is an afterthought introduced by the marketing staff.

American technology transferred to India must develop in a unique way. The transfer of American safety rules and procedures without modification to India is naive. For example, preventive maintenance is not natural to many Third World cultures. Workers are often assigned to jobs without understanding how to respond to nonroutine situations. Often, it is not clear to these workers that safety is as much a function of human interventions as it is of mechanical systems themselves.[31] It is an important concern, therefore, that organizations and individuals in India

[28]*Oregonian*, January 30, 1985.

[29]*Newsweek*, May 26, 1986, advertisement.

[30]B. I. Castleman, "The Double Standards in Industrial Hazards," in J. Ives (ed.), *The Export of Hazard* (Boston: Routledge and Kegan Paul, 1985), pp. 60–89.

[31]G. R. Lanza, "Blind Technology Transfer: The Bhopal Example," *Environment, Science and Technology*, 19(7):581–582 (1985).

act and react differently from those in Union Carbide's plant in Institute, West Virginia. Nevertheless, Union Carbide executives testified that the same safety standards were used in both facilities.[32]

The tendency to assume that other cultures are similar to ours and have the same values seems to be a widely held misconception in the United States. For example, the American military failed to understand the Vietnamese; Americans cannot comprehend why Moslems do not consider democracy the ideal system. The relative importance of O and P perspectives in China differs drastically from that in America and Western Europe. This situation becomes a dilemma when an American corporation effects technology transfer that carries with it the possibility of accidents deemed unacceptable in the host country. The T-focused technical audits of the Union Carbide staff failed to consider vital aspects of Indian culture, such as attitudes toward preventive maintenance and precise adherence to rules of operation. In sum, we see a corporate imbalance of T and O perspectives as well as a failure to integrate them for technology transfer.

Management Short-Term View on Staffing

The losses sustained by UCIL motivated the company to take money-saving steps. The decision to stop refrigerating the MIC storage tank helped lower costs, as did staff reductions among operating personnel. Prior to the accident, the size of the production staff was reduced from twelve to six and the maintenance staff from six to two. One week before the accident the maintenance supervisor position on the second and third shift was also removed and the responsibility added to those of the general production supervisor.[33] The MIC control station had only a single person, although the manual specified the required number as two.

The production supervisor (with the newly added maintenance responsibility) who was on duty the night of December 2–3, 1984, had been transferred from a carbide battery plant only one month earlier and could hardly have had an in-depth familiarity with the operating and newly transferred maintenance procedures.[34]

The T perspective concern for safety is often in conflict with the O perspective concern for profit. This is important to emphasize for it should not be

[32]T. N. Gladwin, "The Bhopal Tragedy," *NYU Business*, 5(1):17–21 (1985).

[33]A. A. DeGrazia, "A Cloud Over Bhopal," *Popular Prakashan*, Bombay, 1985; and V. Morehouse and A. Subramanian, *The Bhopal Tragedy* (New York: Council on International and Public Affairs, 1986).

[34]W. Morehouse and A. Subramanian, *The Bhopal Tragedy* (New York: Council on International and Public Affairs, 1986).

inferred that the T perspective is always totally wrong and the O perspective always totally correct. The shorter time horizon of the O perspective often dominates the longer time horizon of the T perspective with the result that longer term consequences are discounted.

In both the military and industry, job rotation is considered desirable to develop able managers and executives. They should be familiar with all aspects of an organization. In a high-tech environment, this modus operandi raises serious difficulties. Such rotation does not facilitate deep familiarity with a complex operation. By the time a person has learned the details of an operation and its pitfalls, that individual is transferred to another job. At UCIL, a technical manager especially trained at the Union Carbide Institute plant was transferred by UCIL to a non-MIC unit at Madras.

The U.S. military establishment has recognized this dilemma in some exceptional situations. It is exemplified by Admiral Rickover's long association with the Navy's nuclear submarine program. The striking contrast between the Navy and the utility industry nuclear programs thus comes as no surprise.

Organizational Linkages

Hamidia Hospital in Bhopal has very close linkages with UCIL. The company had contributed a wing with sophisticated equipment to the hospital. Many of the doctors have close connections with UCIL executives, which has muted criticisms of UCIL procedures.

The Government

Enforcement of worker safety and environmental rules is the responsibility of state governments. The department of labor in Madhya Pradesh State employs fifteen factory inspectors to monitor more than 8,000 plants. Inspectors in some offices lack typewriters and telephones; they must travel by public bus and train. The Bhopal office has two inspectors, both mechanical engineers with little knowledge of chemical hazards.[35] Even so, some safety lapses were reported to the chief inspector of factories, but he renewed the license annually without acting on them.

The Bhopal development plan, issued August 25, 1975, specifically required plants manufacturing pesticides to be relocated to an industrial zone 15 miles away, yet the existing UCIL plant received an MIC license just two months later.

The Labor Minister opposed moving the plant because it is a Rs.

[35]*Oregonian,* January 31, 1985.

250 million investment and the trade union was against any move. This is a typical conflict of interest characteristic of different O perspectives: the safety of the nearby public versus the welfare of the labor force. It is reminiscent of the arguments in the United States for nuclear power plants—environmental versus economic concerns.

MIC was not a normal factory emission and therefore was not even monitored by the pollution control board. The provincial Labor Minister admits that the state's industrial and environmental safeguards are deficient.[36]

P Perspectives

A Human Limitation:
Ceteris Paribus *Projections*

The human mind has difficulty in dealing with systems that involve multiple interactions that occur simultaneously. This is illustrated in forecasting. We easily focus on one technology and one variable; we then study some of its impacts. Engineers project the evolution of microelectronics and assess future computer capabilities. Demographers forecast future population and look at urban crowding and potential food shortages. But forecasts tend to ignore the fact that everything changes simultaneously. The world does not operate *ceteris paribus*, that is, changing only one thing at a time and leaving everything else unchanged.

We also tend to ignore complex feedback loops in systems. The military sees a threat and develops a new system in response to it, for example, Star Wars. But they then rarely consider how the enemy might effectively counter such a system so that it will operate a minimum rather than maximum effectiveness.

This *ceteris paribus* habit reflects a limitation of the human mind and is conveniently handled by the compartmentalization of problems. It is further justified by the fact that the joint probability of occurrence of two unlikely events is even less than their already low separate probabilities, that is,

$$P(a \ \& \ b) < P(a), P(b)$$

This *ceteris paribus* limitation accounts for the inability of otherwise able analysis to develop meaningful scenarios for the future. It is a source of serious problems in risk management.

[36]*Oregonian*, January 31, 1985.

Another Human Limitation: Misapplying Personal Experience

Discounting the Distant Past and Future

A serious source of error is how the mind summarizes or integrates experience. We focus on two contrary tendencies.

It has long been recognized that we tend to discount the future; the further an event lies ahead, the less we give to its importance. In business, ecology, and health, long-term effects tend to be dismissed relative to immediate ones. The Chinese farmer looks for short-term cash crop benefits; the American business executive focuses on the quarterly profit and loss statement; the smoker on immediate gratification rather than long-term lung cancer. This is not to imply that the discount rate is the same for all. It obviously depends on socioeconomic and cultural factors. The poor discount more heavily than the rich; Americans discount more heavily than Japanese.

The discounting tendency also exists when we look at the past. Tversky and Kahneman[37] have demonstrated this natural tendency in several experiments. In each case, recent past experience is weighted more heavily than the distant past. In integrating our experience, we subtly introduce significant distortions.[38] If no serious problem has arisen in the recent past, we downgrade the likelihood of a future one. Discounting is common to both P and O perspectives. We recall the absence of a long-term view noted under O earlier.

Filtering Out Input Conflicting with Ingrained Views

Perhaps the most powerful barrier to new input is created by our existing mindset. Upbringing and education implant a mindset that proves resistant to change. For instance, a person with strong technical training may be unable to do justice to multiple perspectives.

The tendency to filter out images inconsistent with past experience is one of the major characteristics of the P perspective itself.[39] New evidence tends to be accepted only if it is consistent with one's prior beliefs (developed through education and/or experience) and rejected if it contradicts previously held perceptions. As a result, there may be no active search for information that contradicts accepted beliefs.

The effect of such biases is reflected in the case of Bhopal. Personal physical experience with a hazard tends to magnify its effects in the

---

[37]A. Tversky and D. Kahneman, "Judgment Under Uncertainty: Heuristics and Biases," *Science*, 185:1124–1131 (1974).

[38]H. A. Linstone, *Multiple Perspectives for Decision Making* (New York: North-Holland, 1984), pp. 20–24.

[39]H. A. Linstone, *Multiple Perspectives for Decision Making* (New York: North-Holland, 1984), p. 65.

mind, whereas nonexperience usually minimizes its potential effects. As an old saw has it, "What I don't know can't hurt me."

In an interview after the accident,[40] the Works Manager at UCIL was asked whether he knew that even in fairly low concentrations exposure to MIC could be fatal. He gave "my view, the company's view":

> We do not know of any fatalities either in our plant or in other carbide plants due to MIC. We know that, at 20 ppm concentration, some people found it intolerable, unpleasant to stand around.

Once a hazard is recognized, the nature of its physical manifestation contributes to distortions. Human beings are more fearful of known dangers they cannot readily sense than of those they can easily detect. Again, listen to the Works Manager at UCIL:

> You see, about phosgene, it is much less perceptible than MIC; the symptoms appear after eight hours. So our view is that phosgene is potentially more dangerous.

These comments also remind us of a problem often encountered with the O and P perspectives: They may be difficult to separate without many more in-depth interviews. We cannot be sure, based on the interview, whether the Works Manager gave his own or the company's view.

In most societies, distortions are also caused by the media. In the West sensational news is amplified and unexciting news is downplayed. As a result, the individual overestimates the likelihood of being murdered and underestimates dying of asthma and stroke.[41]

A Worker's Perspective

Human reaction to a fearsome occurrence is intuitive. Suman Dey, a UCIL worker, has recounted the event:

> There was a tremendous sound, a messy boiling sound, underneath the slab, like a cauldron. The whole slab was vibrating. [He started to run away, heard a loud noise behind him, saw the concrete crack and gas shoot out of a tall stack connected to the tank.] I panicked . . . [When the plant superintendent arrived] he came in pretty much of a panic. He said, "What should we do?"[42]

With no senior supervisory personnel on site Sunday night, the oper-

[40]P. Bidwai, "Deadly Delay in Bhopal," *The Times of India*, p. 6 (December 19, 1984).
[41]B. Fischhoff, et al., *Acceptable Risk* (Cambridge, England: Cambridge University Press, 1981); and P. Slovic, et al., *Facts and Fears: Understanding Perceived Risk* (manuscript) (Eugene, OR: Decision Research, Inc., 1981).
[42]*Oregonian*, January 30, 1985.

ators did not activate the alarm, fearing censure for creating public panic.

The operator washing the MIC line did not know about the seepage of water into the MIC tank. Not surprisingly, an empty tank that was kept for pressure relief in case of excess pressure was never used.

The fragmentation and compartmentalization of job responsibilities are another characteristic problem, particularly in areas where the work force has limited training and education. It may be tolerable in a system with loose coupling, but it is dangerous when system couplings are tight.

Ignorance of Outsiders

The residents of nearby squatter settlements were poor and often illiterate. They knew nothing about the plant and its alarm system or MIC and its toxicity. Even the Bhopal chief administrative officer stated that "I have never heard of a compound called MIC," while a factory doctor insisted that the gas is not lethal and only causes eye and lung irritation.[43] Many residents ran toward the plant when the alarm sounded.

An Exception: A Well-Trained Professional

In an adjacent plant, Straw Products Ltd., the General Manager evacuated about 1,400 workers in five buses as soon as the MIC leak was detected. Almost all lives in this plant were saved. This executive had been a Brigadier in the Indian army and his military training provided him with needed crisis management skills.

It should be noted parenthetically that in the Chernobyl accident the thirty or so fires that had broken out were extinguished within 3½ hours after the explosion. In the early stages of the recovery program, intensive weather modification activities were carried out to contain radioactivity. Little rainfall occurred in the area between April 26 and May 31. These actions are evidence of excellent emergency capabilities as well as courage.[44]

The requirements placed on an operator in an emergency situation are in some sense the inverse of those faced in normal working conditions. That person is expected to make correct inferences and decisions about complex phenomena in short times under great stress.[45] It is a situation quite familiar to a military establishment yet a peacetime army's leadership is very different from that needed in wartime.[46]

---

[43]*Oregonian*, January 31, 1985.

[44]*Nuclear Engineering*, October 1986, 31/387, p. 3.

[45]H. J. Otway and R. Misenta, "Some Human Performance Paradoxes of Nuclear Operations," *Futures*, 12(5):340–357 (1980).

The Vast Opportunities for Human Error

The opportunities for human error increase exponentially as the size and complexity of systems grow. The description of the events surrounding the accident under the T perspective suggests some of the realities of the situation. Human errors may be (a) unintentional or (b) intentional. The following are illustrative of experiences throughout the world.

(a) Poor alertness, confusion, illness (mental or physical), drug-caused debilitation, ignorance, misunderstanding (possibly due to language problems), overwork, laziness.

(b) Hostility (e.g., revenge against a superior), bribery or payoff (e.g., to sign off on procedure even if not followed, to accept defective equipment)

There is no way to eliminate these causes even if a large part of the operation is automated.

Integration: T + O + P

The three types of perspective are not independent; they influence each other as shown in Figure 7.1. One example is the design of the processing equipment at Bhopal. It is a product of technologists (T perspective) with some recognition of the U.S. operating environment (O_{US} perspective). This refers to the organizational setting for plants in the United States. In addition, human factors are taken into account in the design as well as the instruction manuals (perspectives of American workers $P_{W,US}$; American supervisors $P_{S,US}$, etc.). As a result, we have:

$$T + O_{US} + P_{W,US} + P_{S,US} + \ldots \tag{1}$$

Even for American installations the integration of such a set of perspectives causes serious problems. We find

- Overconfidence in current technical knowledge.
- Failure to recognize interactions among system components that have been designed relatively independently.
- Failure to anticipate people problems and human responses in crises.

---

[46]H. A. Linstone, *Multiple Perspectives for Decision Making* (New York: North-Holland, 1984), p. 340.

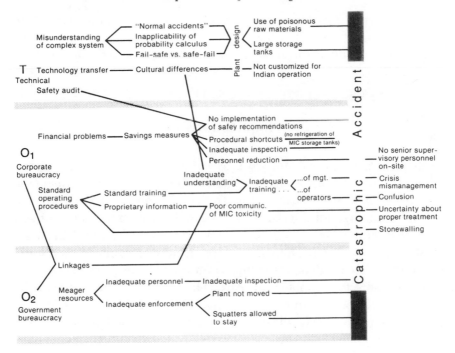

Figure 7.1 Interaction among perspectives—an example

- Failure to anticipate the combinations of unlikely errors with serious consequences (Perrow's normal accidents).

When such a system is transfered into another culture, we have to consider Indian perspectives O_I, $P_{W,I}$, $P_{S,I}$:

$$T + O_I + P_{W,I} + P_{S,I} + \dots \qquad (2)$$

It is unreasonable to expect the integration of equation (1) to be similar to the integration of equation (2). That is,

$$T + O_{US} + P_{W,US} + P_{S,US} + \dots \neq T + O_I + P_{W,I} + P_{S,I} + \dots$$

This suggests a striking weakness in the decision process of multinational corporations.

We see reflections of this situation in other settings as well. The marketing failure of American products in Japan is well known. General Motors (unlike BMW) has exhibited corporate blindness with regard to typical Japanese consumer concerns with quality, size, and product prestige, thus failing miserably in marketing its products in Japan. Johnson &

Johnson learned that Japanese consumers pay much more attention to the packaging of a product than American consumers. It therefore repackaged its products for the Japanese market and, as a result, has been successful.

Implications

Technological fixes will not work as solutions to complex problems. Thus, the first lesson is that *we cannot eliminate fully the possibility of catastrophic failures in complex systems that involve human operators*. Bhopal is neither the first nor the last accident to demonstrate human error as a key contributing factor. Three Mile Island and Chernobyl illustrate similar patterns. Indeed, human and organizational errors account for *at least* 60% of all accidents. It is thus counterproductive for engineers to look invariably to technological fixes. This instills in management the illusion that more technology can solve their problems. To eliminate entirely the possibility of all accidents is unrealistic. Added controls and monitoring systems only increase complexity and introduce new sources of error. Therefore, the first option we should always consider is that of reducing complexity by reorganization or modification of the system itself, not by adding to it.

Seek Means To Reduce the Likelihood of Catastrophic Consequences with Given Systems

Some obvious steps may be deduced from examination of several perspectives:

- T: Try to make the system "safe-fail" by *decoupling* subsystems so that an accident can be bounded or limited to one subsystem.
- O: *Transform* instruction manuals to allow for *cultural differences*, resulting in effective equivalences in practice, rather than merely relying on literal equivalence in language.
- P: Include individuals in the *early stages* of planning overseas operations who have the necessary perspectives to balance the likely dominance of T and O.

The following steps entail the agreement of local management. Since such management is often required to be composed of native personnel by the host government, their distinct values come into play and pose unique challenges in implementing the proposed steps:

- O: Insist on crisis management training for *all* managerial personnel or create special crisis management teams as in a growing number of American corporations such as United Airlines and Dow Chemical.[47] This splits operations into routine and crisis conditions, with distinct expertise required for each; the crisis teams must be trained to be especially alert for early warning signals such as minor incidents that can warn of more serious accidents. They also need to be familiar with normal operations so that they can appreciate the transitions required to go from normal to crisis conditions.
- O: Improve the training of workers, again keeping in mind cultural differences that alter meaning, learning, discipline, values, etc.
- O: Recognize that Third World governments do not have the resources to adequately monitor compliance with their own safety rules and controls; therefore accept the burden of corporate responsibility to augment the level of local plant monitoring to fill this gap.
- O: Institutionalize risk assessment and safety function in the company, clearly specify its responsibility and authority, and require it to report directly to the local top management, short-circuiting the bureaucratic hierarchy.
- P: Give investigative journalists and "whistleblowers" more protection in exposing poor practices, thus anticipating potential catastrophes.

Persons with backgrounds other than science or engineering are likely to have a different balance (or imbalance) in perspectives. In particular, lawyers, career bureaucrats, and journalists are more likely to be attuned or sensitive to such factors as:

- O: Cultural differences relevant to technology transfer; proprietary nature of relevant information; stonewalling in the face of disaster ("circling the wagons"); government and corporate bureaucracy—ignorance, obsolete regulations and SOP's, nonenforcement, political pressure, suppression of uncertainties organization linkages.
- P: Unintentional errors by workers; intentional failures.

The journalist can effect communication when the engineer/scientist is stymied; the lawyer can circumvent proprietary concerns. The bureaucrat can pinpoint bureaucratic maneuvers that often escape the detection of scientists and engineers and figure out how to make a safety audit group effective in a large, disinterested organization, for example, having it autonomous, reporting directly to top management.

Implementation of all these steps, difficult at best, is no absolute *guarantor* of the prevention of major catastrophes. There is another path.

[47]"Coping with Catastrophe," *Time,* p. 53 (February 24, 1986).

Redesign the Existing System To Reduce Dangers

It is often possible to redesign the existing system to reduce dangers further. This suggests an altered design philosophy giving much greater weight to:

- T: decoupling of subsystems as suggested above—a task easier in the initial design phase than later stages.
- O/P: nontechnical sources of accidents and greater balance between the technical and human aspects of design.
- O/P: the unique requirements of use in other cultural environments; this means that systems must be customized to the needs of other cultures to increase safety.

This calls for early introduction of multiple perspectives in the design phase. Although this still cannot assure complete elimination of the possibility of catastrophic accidents, redesign can respond to anticipated situations, including some events previously thought inconceivable. But not all events can ever be anticipated and new possibilities may arise from the redesign itself. Recall the lessons learned from computer programming: *no general program exists that can correct the errors of any program including itself.* This leaves us with the paradoxical situation that the human operator is most needed when a completely unanticipated situation is at hand, "the condition when he [or she] is least able to respond in a constructively creative way."[48]

Probe Conceptually Different System Solutions That Avoid Catastrophic Consequences Altogether

Common sense dictates that we search for alternative systems in which failure simply cannot lead to unacceptably catastrophic consequences. An example is the Bayer facility in Dormagen, West Germany. It has produced MIC for twenty years using nonpoisonous raw materials.[49] Thus:

- T: alter the production process or use materials in such a way that the same need can be met in a technologically new way, one that excludes catastrophic accidents.

This constitutes a challenge to the technological community and an impetus for innovation. In the field of energy, we clearly have alter-

[48]H. J. Otway and R. Misenta, "Some Human Performance Paradoxes of Nuclear Operations," *Futures,* 12(5):340–357 (1980), p. 355.

[49]S. K. Subramanian, "Die Bhopal-Katastrophe," *Echo,* April 1986, Carl Duisberg Gesellschaft, Koln, F. R. Germany.

natives to nuclear energy. We expect that conceptually different system solutions will become available, for example, solar and fusion systems. But suppose we fail. Then . . .

Ask the Final Question

A societal decision is now unavoidable: *Can the consequences of a catastrophic accident be tolerated?* History has shown that the human and ecological resilience to disaster is enormous. But there are limits and this is no excuse. To date, industrial catastrophes such as Bhopal and Chernobyl have been "tolerated." Yet, the evolution of technology drives the growth of system complexity and is inevitably accompanied by two disturbing features:

1. Possible kinds of catastrophic accidents may grow. Example: AIDS casts the shadow of the possibility of a future bioengineering error creating a deadly virus that sweeps over the world with lightning speed. So far, this is a figment of the novelist's imagination (e.g., *The Andromeda Strain*), but it seems far less remote than it did twenty years ago. In the field of chemistry, new chemicals have been registered at the astonishing rate of seventy per hour in one recent year.[50]
2. The demands on management capability become more severe. Example: America's national strategic command and control capability is already seriously constrained by the lack of expertise, or illiteracy, in electronic technology on the part of responsible military commanders who must make critical decisions. According to a recent Undersecretary of Defense, the Pentagon has been unable to properly manage the command and control program. And there is increasing concern about the ever greater pressure on the ability of a president to manage a nuclear crisis. He may well be a figurehead, an "honorary commander-in-chief."[51]

[50]*New York Times*, December 16, 1984.
[51]D. Ford, *The Button: The Pentagon's Strategic Command and Control System* (New York: Simon and Schuster, 1985), pp. 121, 186, 227.

CHAPTER 8

Paradigm Breaking: Surfacing and Challenging Key Assumptions

What American industry has to learn from the Japanese is not to be learned through inflating the value of their example. The proper lesson is of a different sort. What the Japanese have done is to build an approach to the work of manufacturing that takes explicitly and centrally into account the realities of the new industrial competition. By contrast, American managers too often view their work through a haze of outdated *assumptions* [emphasis ours] and expectations.

> William J. Abernathy, Kim B. Clark, and Alan M. Kantrow
> *Industrial Renaissance, Producing a Competitive Future for America*
> (New York: Basic Books, 1983)

Throughout this book, we have continually stressed the need to challenge key assumptions. This perpetual challenging is one of the most critical and central aspects of New Thinking. For this reason, we devote this entire chapter to the process of surfacing and challenging key assumptions. To make its understanding as clear and direct as possible, we use an actual case to illustrate the process.

The Case of the Drug Company

Imagine that you are the chief executive officer, or CEO, of a pharmaceutical company and are faced with a problem that threatens in one blow to wipe out your entire business—certainly an attention-getting situation.

The problem is this (and as we will see, it has several levels; unlike exercises, real problems only unfold over time): you produce a product, a painkiller, which, because it has a narcotic base, can only be obtained by prescription through a physician. After receiving a prescription for your drug, a well-known brand-name product, the patient takes it to a pharmacist. The pharmacist in turn either fills the prescription as instructed by the physician without comment, or says something to the patient like, "There is a generic-brand substitute available at a much lower cost than the brand name. If you like, I can give you the cheaper generic brand. Which do you prefer?" In some states, the pharmacist is required by law to inform the patient that a generic substitute is available.

While potentially beneficial to the patient (assuming that the generic-brand drug is of equal quality), this action is a potential financial disaster for the drug company. It threatens in one blow to wipe out one of its mainstay products. Since the drug annually generates millions of dollars, the company's entire financial structure is threatened. As CEO, what do you do?

You can do several things. You can attempt to think the problem through yourself, taking the whole burden on yourself, either by not delegating it to anyone else or not entrusting it to them. You can then choose, on your own, the best option to combat the threat. Or you can do something else which, depending on how it is done, can either replace or supplement the alternative of proceeding by yourself. You can choose to involve others in analyzing the problem. Since it threatens financial disaster for the entire organization, it might be desirable to involve others in a consideration of their own fate and for the good of the whole. Also, more heads may be better than one in coming up with needed and creative alternatives, especially in a crisis situation when critical faculties are likely to be blunted and stressed. No single mind—except God's—can ever know all there is to know about any complex problem. (God's is the only mind that can constitute a single, self-contained guarantor.) The final decision will still be yours, but if you have the right style or corporate culture, this need not preclude participation by others in the analysis of the problem and perhaps even in the final decision. You would certainly like others to go along with whatever you decide, not to sabotage it.

In this authentic case, the CEO decided to involve some of his senior executives in the analysis of the problem. The CEO felt he had no choice but to solicit the most widespread expert advice from as many different parts of the organization as possible. As a consequence, twelve key executives representing all aspects of the business were assembled. Here is precisely where the deeper and more interesting aspects of the problem began to emerge. What happened is a powerful illustration of how real problems only unfold over time.

A strange thing occurred. The twelve executives split themselves into three subfactions. This was not done out of any animosity, but because complex problems naturally suggest more than one "best" alternative. The senior executives began to coalesce around the particular alternative that individually made the best sense to them. Each group then proceeded to make the strongest case for its alternative to the exclusion of the other two.

The three alternatives were as follows: the first group wanted to lower the price of the drug; the second wanted to raise it; and the third wanted to keep it the same before the crisis hit. The first group in effect wanted to "out-generic" the generic drug by making the company's drug into a generic, at least in terms of price. This alternative or policy is the one that people most readily think of. It is in many ways a defensive policy, which doesn't necessarily make it "wrong," for in complex problems there are not always clear right or wrong answers. Some responses may be stronger than others along some grounds, but very rarely is one alternative strongest or best on all dimensions.

The second alternative is the diametric opposite of the first. Furthermore, it shows that in the realm of complex problems not only is there usually more than one serious alternative, but, further still, generally at least *one pair* of alternatives exist such that *each* member of the pair is the dialectical opposite of the other. Thus, in this case, alternative two is the dialectical opposite of alternative one.

The reasoning behind alternative two is even more fascinating and important for our purposes. The group of executives supporting this alternative argued that, faced with the threat of generic drugs, they had to do something that would communicate to the marketplace the difference between their drug and others. This group felt that by raising the price of the drug they would be displaying increased confidence in their product's quality to the marketplace. They were thus making an important and largely unstated assumption about consumer psychology: some people will buy a product if its perceived quality, as signaled through a greater price, is high.

The last group argued something entirely different. While the first two were oriented toward the external marketplace and the price consumers would be willing to pay, the last group was oriented toward cutting internal costs of production. They argued that if the price of the drug were maintained at current levels or, at the very least, set midway between the proposals of the first two groups, then they could raise profits by cutting internal costs. They proposed to do this by eliminating the Research and Development (R&D) arm of the company, the largest source of internal costs to the company and to drug companies in general. Their argument was that if the current price of the drug was sufficient to generate necessary revenue for the company, that is, if the demand for the drug remained stable, then the company would not need to develop new products. To say the least, the R&D department as a critical stakeholder or partner in the company wouldn't be overjoyed with this, but then "business was business," or at least as this group saw it.

Since each group was of roughly equal power in the organization, no single group could force through its pet alternative over the objections of the others. Each group had to convince all the groups if the company was to embark on a unified course of action that everyone could embrace with confidence.

How then did each group try to persuade one another of the correctness of their individual policy? They did what most managers and executives have been trained to do. They analyzed past data, for example, past sales volume (amount sold) for various selling prices of the drug, and, where they could, they collected new data, for example, from trade magazines and reports from salespeople from the field. That is, they used methods based on the philosophical IS's described in Chapters 2 and 3 to settle the issue. The trouble was, in this case at least, the data didn't settle anything and actually made things worse. They only convinced the proponents even more of the truth of their separate policies. The reason may be the most illuminating and instructive aspect of the entire problem.

One of the valuable things we learn in school is to test our ideas against the criticisms of others and, wherever possible, challenge data from the outside world. In school this tactic generally succeeds because, as we pointed out previously, the kinds of problems that are presented to students are so greatly simplified that they are exercises and not really problems in the true sense at all.

In this case, more data merely confounded the "mess" management was faced with. Indeed "mess" is a more appropriate word to use in describing this case than the more benign word "problem." Russell Ackoff[1] defines a "mess" as a *system of interacting problems*, none of which can be formulated independently, let alone solved, independently of all the other problems on which it impacts and which impact on it. In addition, each group was *assuming* different things about the nature of the problem. Each group was taking certain things for granted—as "true"—without conscious or explicit knowledge that they were doing this.

As a result, each group was selectively reinterpreting the data they had in common, unconsciously of course, to "prove" its particular case.

[1]Russell L. Ackoff, *Redesigning the Future* (New York: John Wiley, 1974).

Further, where common data from the past were not available, and hence where each group had to collect new data, *each group was collecting different data from different sources. Each source was designed, again largely unconsciously, to prove each group's individual case.* Hence, instead of data really testing each alternative, each alternative was being used to direct its believers into procuring data that would confirm the validity of its alternative. A very circular process indeed, and one from which management had tried repeatedly, without success, to extricate itself. But since everything depended on which assumptions were made, as in fact every complex problem does, the case couldn't be analyzed without some important critical assumptions. But since very few, if any, of those assumptions were ever raised to the surface for conscious inspection and challenge, each group cycled around its own vicious circle.

(In the spirit of Chapter 4, each group was using a different model to collect different data and interpret common data selectively. Thus, each group was using a different Multiple Realities IS.)

Not that management didn't try to break out of this circle. They used every financial model (Analysis) and approach they were aware of in an attempt to get some neutral piece of data (Agreement) or critical finding that would once and for all clearly differentiate between them. However much they tried, their expectations were consistently dashed.

At this point Mitroff and a colleague, James Emshoff, entered the scene in the role of consultants. For years, Emshoff and Mitroff had worked on problems of this kind independently of one another. A year's visiting appointment for Mitroff at the Wharton School of Finance, University of Pennsylvania, gave them the opportunity to combine their insights. As a result of this collaboration, Emshoff and Mitroff achieved a "breakthrough" (a much overused word) in formulating a practical and operational method for handling complex, messy problems. Since the method has been described extensively elsewhere, we merely review it here.[2]

Assumptional Analysis

It soon became clear to Emshoff and Mitroff that fundamental differences in basic assumptions were at the heart of the disagreements. Each group was making a fundamentally different set of assumptions about the "real nature of the problem." No wonder more data didn't settle anything—it only served to activate underlying differences. It

[2]Richard O. Mason and Ian I. Mitroff, *Challenging Strategic Planning Assumptions* (New York: John Wiley, 1981).

didn't test or resolve them. It only made things worse. We have a perfect example of where "more can lead to less." Since for the most part the assumptions remained buried and implicit, the groups themselves were largely unaware of what was happening. All they knew was that time and again they disagreed and were frustrated.

In effect, a dialectical standoff existed between the groups. While each group constituted a separate Multiple Realities IS, collectively they constituted a Dialectical IS. The difficulties were that the groups didn't know this and, hence, tried to use agreement to break their impasse. Agreement, however, was, by definition, unable to handle Conflict.

How could we get the participants to reveal to us and to themselves in a nonthreatening way the underlying assumptions driving their different "solutions"? Like new thinking, assumptions are like the weather. Everyone talks about them, their influence, and their crucial importance. Beyond this, no one offers a way of getting at them or dealing with them. This was our goal.

Both Emshoff and Mitroff had for some time been working with a concept that was a bare stone's throw away from assumptions—the concept of stakeholders. Stakeholders are any individual, group, organization, institution that can *affect* as well as *be affected by* an individual's, group's, organization's, or institution's *policy* or *policies*. Figure 8.1 shows this in a schematic way. Notice that a double line of influence extends from each stakeholder to the organization's policy or policies and back

Figure 8.1 A stakeholder map of the drug company

again—*an organization is the entire set of relationships it has with itself and its stakeholders.* An organization is not a physical "thing" per se but a series of social and institutional relationships between a wide series of parties. As these relationships change over time, the organization itself changes. It becomes a different company. The failure to grasp this has prevented many an organization from seeing that it is not the same because its environment, that is, its external stakeholders, has changed even though internally it looks the same. Since we are dealing with a system, a change in any one part potentially affects all other parts and the whole system itself.[3]

Let's examine Figure 8.1 a bit further to see its implications. Every organization has some form of external competition. While not strictly a formal part of the internal organization, it nonetheless affects the organization and its policies. Every organization must continually ask itself such questions as: "If we do such and such, what will our competition do? If we enter this market, will our competition grow, leave, retaliate? Who is our current competition? How determined and how strong are they? Who might our future competition be? Can we prevent our competition from entering our turf, our market in the first place? Can we raise the 'entry costs' (e.g., capital accumulation) high enough to keep them out or at least make their entry difficult and costly? If we decide for whatever reasons to get out of a particular market, can we do it so that we make it more difficult for our competition to get out of a worsening situation? If we get out first, will it be more difficult for others to get out later?" These questions are so important that an exciting new school of business policy and planning has arisen to consider them explicitly.[4]

Competition is certainly an important factor in the present case since it was the explicit competitive threat from lower-price generic drugs that started the whole crisis. But so are all the other stakeholder parties. The company's sales force is an important stakeholder since whatever policy the company enacts potentially affects their commissions and hence their motivation to sell the drug. Physicians and pharmacists are also obviously important since they and not the drug company have direct contact with the customers, the patient. Their behavior—that is, their attitude toward the company and its products—is obviously an important factor in the patient's behavior.

Consider a few more stakeholders. Government is an important stakeholder in this case in at least two important roles. First, it regulates through the Federal Drug Administration, the sale, release, testing, use,

---

[3] Ackoff, *op. cit.*, and Mason and Mitroff, *op. cit.*

[4] Michael Porter, *Competitive Strategy, Techniques for Analyzing Industries and Competitors* (New York: The Free Press, 1980).

and distribution of drugs. This is especially true where narcotics are involved. Second, since the drug has a significant narcotic base, the government comes into the picture through the procurement and regulation of opiates from foreign countries.

Finally, consider the stakeholder, the holding company. The drug company in this case was owned by another larger pharmaceutical company "50 miles up the road," so to speak. They certainly had a stake in whatever the subsidiary decided to do for it would certainly affect the profits of the larger parent company.

How does all this involve assumptions? Simply put, *assumptions are the properties of stakeholders.* The proponents of the different policies in the drug company were disagreeing, often strongly, because they were positing very different properties about the behavior of the stakeholders. No one had the definitive data, information, and arguments to know beyond all doubt what all the stakeholders were like or how they were likely to behave in all situations.

The bigger, the more complex the problem, the more it is likely to involve a wider array of stakeholders. As a result, the more assumptions needed. It is a characteristic and fundamental feature of complex problems that not everything of basic importance can be known prior to working on the problem. *Rarely do we have a clear statement or definition of the problem before we begin working on it.* Rather, such a statement often only emerges with difficulty over time and only as a direct result of our working on it. The definition of important problems entails intense human creations, not made-up situations that come from textbooks.[5]

At last we come to the crux of the problem. Consider the single stakeholder, the physician and the accompanying assumptions. For ease of presentation, we will consider only the two groups, the one wanting to raise the price of the drug and the one wishing to lower it. The group advocating a price increase was assuming implicitly that physicians were motivated primarily by the traditional model of medical care: physicians were concerned primarily with the health and well-being of the patient *irrespective of cost.* This group was assuming that physicians were *price-insensitive* about the cost of the drug—they would prescribe the drug if they were convinced of its quality. A further assumption was that the physician's recommendation would be critical in overcoming the counter-suggestion of the pharmacist. As one can see, a whole bundle of assumptions are actually tied up with the physician and his or her effect on other stakeholders such as the patient and pharmacist.

On the other hand, the group that wanted to lower the price was

[5]Ackoff, *op. cit.*

assuming implicitly that because of the skyrocketing cost of medical care, physicians were becoming increasingly *price-sensitive*. They would no longer prescribe a drug or treatment merely because of its quality regardless of its cost. At some point, quality would have to give way to cost.

Emshoff and Mitroff introduced a special wrinkle that made it possible for the different groups to see their assumptions, let alone their effects. This was a simple, yet effective, way of mapping or plotting assumptions, shown in Figure 8.2.

Once pertinent stakeholders have been identified and the assumptions associated with them have been surfaced, some assumptions are usually more critical or important to the success or viability of a policy than others. One also feels more confident about the truth or certainty of some assumptions than others. Thus, Figure 8.2 demonstrates that all the groups, but for very different reasons, regarded the physician as the most important but uncertain stakeholder. Consider again the high-price group and understand that it is the *assumption* about the stakeholder, *not* the stakeholder itself, being plotted in Figure 8.2. Thus, the assumption that "Physicians are price-*in*sensitive" is both the most critical (important) assumption to the success of the high-price group's policy, but, when stated, is also the *most uncertain*. That is, *the assumption is as likely to be true as false.* (Greatest uncertainty exists when an assumption is 50–50. For instance, greatest uncertainty is present if rain is just as likely as sunshine, i.e., there is a 50% probability of rain. On the other

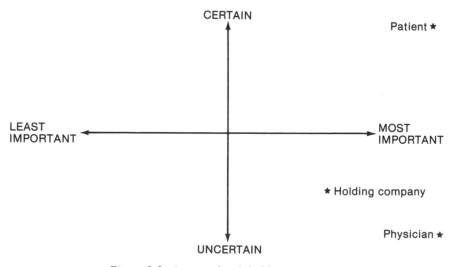

Figure 8.2 A map of stakeholder assumptions

hand, if the chance of rain is 100% [and thus the chance of sunshine is 0%], or if the chance of sunshine is 100%, then one has greatest certainty. One knows *for sure* whether it will rain or will not.)

Without the assumption of price insensitivity, this group's policy cannot fly. At the same time, it is the most open to doubt. Are *all* physicians price-insensitive? Or merely some? If so, what are the "some" who are price-insensitive like? What are their personal and demographic characteristics? No wonder more data or Agreement didn't settle anything because the data that were relevant to these questions were either unavailable—they had never been collected—or were ambiguous. One could infer whatever one wanted from the data because the available data were all "mixed up." They contained data on physicians of all kinds.

The second most important and uncertain stakeholder was the holding company. If you go with raising the price, then you have to assume, as the high-price group did, that the holding company is primarily interested in maximizing profits. It will sell less of the drug (lower volume), but make more per drug sold. On the other hand, if you lower the price, you have to assume, as the low-price group did, that the holding company is primarily interested in maximizing market-share. You sell more of the drug (greater volume), but you make less per drug sold. You assume that the holding company is primarily interested in becoming the market-leader in its industry.

Neither assumption is necessarily wrong; they are merely different. Which is best depends on the overall objectives of the parent company and the subsidiary drug company considered as a total system. *Notice that although we started with just one part of the system, the drug company, we have ended up having to consider the whole system. This is in itself one of the prime features of complex problems. They quickly involve or turn into a whole systems problem.*[6] This recognition is in fact the core of Unbounded Systems Thinking.

If the assumption involving the parent holding company was so important, one might ask, "Why didn't the drug company merely call up or visit its 'parent' so close by?" The answer is, "If only organization politics and jockeying for power, prestige, and influence were that simple." Just to ask for clarification of something in some systems is not as clear-cut as it may appear. Nor is it necessarily rewarded by those "on top" who may not know the answers themselves and as a result do not want to reveal their ignorance. In short, the culture of the organization may have precluded such requests.

[6]Ackoff, *op. cit.*

All the groups, on the other hand, felt that the patient wanted a low-cost, quality product. Hence, all the groups placed their assumption regarding the patient in the upper right quadrant.

We will not bother to pursue here the further details of our methodology for resolving messy problems of this kind.[7] We merely note that the group of executives was finally able as a result of this process to agree as a whole on one of the alternatives. They decided to raise the price of the drug in certain key, test locations and monitor very carefully the reactions of critical stakeholders, and, in this way, test the validity of selective, crucial assumptions. If one lowers the price of the drug when one could have raised it, one will never find this out from the market's response or behavior. One will have precluded the opportunity of discovering this. On the other hand, if one raises it in certain test locations, one will learn very quickly if the market will tolerate this action.

Note that one can still continue to disagree with the company's final action from a variety of other perspectives, one of which at some point would certainly be ethical. Our emphasis would nevertheless be that in either agreeing or disagreeing with the company, one would still be making some fundamental assumptions. There is no way to avoid making assumptions of some kind. Therefore, they must be displayed and examined in such a way that they can be debated. *Far better to debate a question without necessarily settling it than to settle a question without debating it.* While we understand and empathize with the need for executives to take decisive action, it is better to delay than to take the wrong actions for the sake of expediency.

Review of the Process

Let us briefly review some key features of the process we have been examining. We started out by explaining that assumptions are rooted in the behavior of someone or *some party;* assumptions, in short, pertain to stakeholders; they do not exist in a vacuum. Generating stakeholders is a concrete way of getting at assumptions. Most persons cannot generate assumptions directly. They are too vague, too hazy, too hidden from view. Asking people, on the other hand, to list the set of actors, parties, and so on, who are affected by one's actions is both a concrete and easily accomplished task. We have never encountered an organization that couldn't do it, *if* its culture was conducive to it. Once stakeholders are identified, it is then a relatively easy step to ask, "What do I have to assume is 'true' of a particular stakeholder (i.e., its behavior) such that

---

[7]Mason and Mitroff, *op. cit.*

starting from this assumption I can then derive or support my policy or actions?"

Note that there is no guarantee, or guarantor, that all groups will generate the same set of stakeholders as happened in the drug company case. This is thus one of the first ways in which groups can differ. If they do not generate the same set, then they are making different assumptions about who is influencing or who ought to influence (at the very least be considered) in their situation. Many a group differs over "the basic right of recognition." As the modern corporation has grown, it has had to consider more and more stakeholders than it previously did, whether it likes this or not. The same is now true of all organizations, public and private.

The second way in which groups can differ is in the qualitative form of the assumption or property they impute to a particular stakeholder. The physician in the drug company case is illustrative. One group assumed price-insensitivity; the other assumed precisely the opposite. Thus, the process helps to surface fundamental differences of this nature.

Third, groups can disagree over their importance and rankings. Thus, for example, two groups could conceivably both agree that the physician was a relevant stakeholder. They could even both agree on the same qualitative assumption of price-insensitivity. However, one group could consider this to be very important and very uncertain to their policy while another could regard it very unimportant and very certain. The step of mapping assumptions helps groups to see this.

We have deliberately emphasized the word "see" because we believe strongly that often it is the failure to be able to observe assumptions that drives groups around in an endless circle. We have also deliberately emphasized each step in the process of surfacing assumptions because "a method" for working on complex problems is above all a behavioral process for allowing persons to see their differences: over stakeholders (who's involved, who should be considered, who has the right to be recognized), over assumptions (what the stakeholders are presumed to be like), and over mapping (what's important and what's felt to be known). Many difficulties between our fellow humans are as much due to the methods we use to examine our differences as they are the result of basic differences themselves. We relish organizing ourselves into forms and forums that are designed to prevent us from airing and resolving our differences in anything resembling a constructive fashion. We must learn to break out of this vicious and destructive circle.[8]

[8]Ian I. Mitroff, Richard O. Mason, and Vincent P. Barabba, *The 1980 Census: Policy Making amid Turbulence* (Lexington, MA: Lexington Press, 1983).

Conclusion

This entire book has attempted to introduce a new way of thinking about complex problems and social systems. In particular, this chapter has argued that social systems are structured, organized, and constituted in terms of stakeholders. It has also argued that we need new methods for systematically uncovering important stakeholders and their associated properties (assumptions) on which every organization's plans and actions depend. Again, we have deliberately emphasized the word "systematically," a key word. Many organizations pretend to do what we have advocated in this chapter, but, on deeper examination, they really don't. They certainly don't follow the thorough and systematic job of monitoring stakeholders as we are advocating here.[9] Little wonder why their policies stagnate and atrophy as the world changes.

Likewise, it is easy to identify four stances, really pathologies, toward planning that often develop in each of the four cells in Figure 8.2. The authors have encountered groups that felt all their assumptions fell into the unimportant, but certain quadrant. This group is saying in effect, "We know it all (everything is certain) but it doesn't matter (it's all unimportant.)" Now, either this group is right or they are playing a gigantic game of denial. The authors tend to believe it is denial (they are denying the actual uncertainties in their environment), or they are not risking any significantly new ideas or products that would inevitably get them into the uncertain cell.

Another attitude is to say that you don't know anything for certain (it's all uncertain) and you also know that it's all unimportant! In other words, *you know that you don't know but you also know that it doesn't matter.* This can be another defensive position. Again, we are talking about if *every* assumption a group makes or the vast majority of their assumptions fall into the unimportant, uncertain camp.

Another position is to say that one knows it all and that it's all important. In other words, the group is really saying that they're in total control. Now this is either true or very arrogant, or both. When anything falls into one quadrant exclusively, the authors tend to be skeptical. We find this very hard to believe. So do other groups who are listening to this group's presentation of their assumption map.

Finally, there is the attitude that everything is important but uncertain. These groups tend to feel overwhelmed by the world, if not experiencing chaos.

---

[9]See also Ackoff, *op. cit.* for another approach and argument for the *systematic* monitoring of stakeholders.

All these extreme groups need help. Whenever an orientation is so extreme, so one-sided, the danger is that it will fail precisely where it has ignored the concerns that its counterbalancing opposites raise. But this is precisely why the authors advocate the creation of opposing dialectical perspectives on problems of critical, major importance to an organization. If major differences in group perspectives do not naturally exist within the organization, as they did in the drug company, then they must be created by design if need be to ensure that important dimensions and policy options of a problem will not be systematically overlooked.[10]

Too much rides on today's problems to pursue them from one and only one perspective, no matter who the advocate for any one position may be. The most critical thing we can do is to examine problems systematically from several different perspectives.

Postscript

In the years since Mitroff and Emshoff first worked with the drug company, we and our colleagues have had occasion to apply the method described in this chapter to numerous organizations, both public and private.[11] We have yet to encounter an instance wherein an organization did not profit significantly from the use of the technique as long as the organization's culture was supportive of examining its problems. This does not imply that the method is perfect.

However many applications one subsequently performs, there is probably always a certain fondness for one's first application. One can probably never recapture fully the excitement of learning and doing something new for the first time.

In the years since Mitroff and Emshoff first worked on the drug case, we have talked about it many times in making presentations to various organizations. It still remains one of the clearest and cleanest examples of the method.

In all those years we never once mentioned the name of the drug company. While the case doesn't disclose anything really proprietary about the company, and while we weren't prohibited by the organization from mentioning its name, we didn't see any point in naming it.

Then, a major corporate crisis occurred that proved to a landmark event, the injection of cyanide into Tylenol capsules.[12] The drug com-

---

[10]See Mason and Mitroff, *op. cit.*, for methods on how to do this.

[11]See Mitroff, Mason, and Barabba, *op. cit.*; Ralph H. Kilmann, *Corporate Culturing: Getting America Out of Its Rut* (San Francisco: Jossey-Bass, 1984).

[12]See Ian I. Mitroff and Ralph H. Kilmann, *Corporate Tragedies* (New York: Prayer, 1984).

pany is McNeil, makers of Tylenol. The parent holding company is Johnson and Johnson. Perhaps if they had used stakeholder analysis regularly, then they might have anticipated, and responded better, to the families of the victims who recently sued McNeil for its failure to take more effective preventive actions to safeguard their product.

PART V

THE FUTURE OF KNOWING

CHAPTER
9

Conclusion:
To See The World as an
Interconnected Whole

Three propositions:

1) God is all powerful.
2) God is all good.
3) Terrible things happen.

. . . the dilemma has always been this: you can match any two of [these] propositions, but never match all three.

The historian Jeffrey Burton Russell asked, "What kind of God is this? Any decent religion must face the question squarely, and no answer is credible that cannot be given in the presence of dying children." Can one propose a God who is partly evil? Elie Wiesel, who was in Auschwitz as a child, suggests that perhaps God has "retracted himself" in the matter of evil. Wiesel has written, "God is in exile, but every individual, if he strives hard enough, can redeem mankind, and even God himself."

Perhaps God has other things on his mind. Perhaps man is to God as the animals of the earth are to man—picturesque, interesting and even nourishing. Man is, on the whole, a catastrophe to the animals. Maybe God is a catastrophe to man in the same way. Can it be that God visits evils upon the world not out of perversity or desire to harm, but because our suffering is a byproduct of his needs? This could be one reason why almost all theodicies has about them a pathetic quality and seem sometimes undignified exertions of the mind.

Lance Morrow
"Evil," *Time*
June 10, 1991, pp. 51–52

We have argued throughout that we cannot hope to find solutions to our problems if we persist in our old ways of thinking. The new thinking we must develop is encapsulated in Unbounded Systems Thinking. UST contains, but is not completely synonymous with, such tools as the Multiple Perspective Method and Assumptional Analysis. All are intended to improve the ways we deal with complex problems.

In this chapter, we want to examine the fundamental role of ethics in the design of systems for the Information Age. In the seventeenth and eighteenth centuries, Western scientists saw ethical issues as central to science yet, for reasons that are still unclear, the intellectual community turned away from these issues in the nineteenth and twentieth centuries.[1] (We suspect this was because ethics is not susceptible to the methods of either Agreement or Analysis.) UST draws them back in as central and critical features of its approach. Indeed, the question, "How can we strengthen the ethical basis of all our so-called solutions to large-scale systems problems?" is one of the most central concerns of UST. One way, although it is not entirely satisfactory, is through the pursuit of ideals.

The Reality of Ideals

Although far from recognizing and admitting it, every philosophic system contains Ideals. Indeed, by definition, every system must contain them. Each philosophic system posits certain fundamental entities or desired end states of nature or humankind that are unachievable within any finite time period.

Ideals are fundamentally different in principle from goals or objectives. Goals and objectives are desired end conditions that one either achieves, or does not, within some *finite time period.* For instance, company X's may be to make a 10% return on investment over the coming year. At the end of the year, one then measures how much of one's goal or goals one has achieved.

Ideals, on the other hand, such as "peace on earth" or "the feeding of every hungry person on the planet," may not be achievable within any finite time period. This does not make them useless, for the fundamental purpose of Ideals is to urge humankind on in quest of a better end state.

Ideals thus serve as critical benchmarks. They are desired ends that one, it is hoped, approaches indefinitely even if one can never achieve them completely.

Every philosophy is measured not only by what it achieves, but by its aspirations. For instance, the attainment of complete or perfect Truth as posited by Analysis and Agreement is itself an Ideal. The fundamental measures of performance of both Analysis and Agreement, which they take as objectives, are instead Ideals.

[1]C. West Churchman, *Thought and Wisdom* (Seaside, CA: Intersystems Publications, 1982).

Ideals represent the highest aims and aspirations of humankind. They function not only as intellectual, moral, and aesthetic guideposts, but more deeply, as spiritual aides. Ideals not only represent the spirit of humankind writ large, but, even better, writ whole.

To formulate appropriate Ideals, it is helpful to review earlier versions. Earlier formulations were not only the "children of their times," but as such, contained both the strengths and weaknesses of their parents. They were not only deficient, but these deficiencies arose from inherent weaknesses that cannot be rectified through surface changes alone. Contemporary crises such as Bhopal and Exxon Valdez not only pose serious threats to earlier formulations of humankind's Ideals, but necessitate their serious reformulation.

The reformulation of Ideals not only is an expression and recognition of some of the extreme negative effects we are bequeathing to future generations, but also an attempt to pass on a more positive gift, the gift of hope, the reorientation of our civilization to a culture of respect for the environment, not one of blasphemous destruction. This reformulation is anything but a trivial change in words. It signals a profound change in spirit.

Two of Russell Ackoff's earlier books, *Scientific Method,*[2] and *The Design of Social Research,*[3] contain a succinct and especially important statement of the nineteenth and earlier twentieth century's concept of Ideals. Ackoff identifies four primary Ideals that in one way or another have guided Western civilization: (1) the scientific Ideal of perfect knowledge; (2) the economic Ideal of plenty or abundance; (3) the moral-ethical Ideal of goodness; and (4) the aesthetic Ideal of beauty.

From the perspective of this book, the earlier Ideal of knowledge needs to be reformulated to read *humane* knowledge instead of *perfect*. Knowledge not only must be put in the service of humankind as a whole, but more fundamentally in the service of the entire environment. To accomplish this requires that the former economic Ideal of plenty or abundance be recast into an Ideal of "harmony" or "development," and not that of "growth" per se. We must recognize that on every front of our existence "less is now more." We should have learned by now that the unrestricted pursuit of plenty or growth invariably leads to excess. (Many such as Donald Trump may be fundamentally incapable of learning this.)

The Ideal of goodness has to be expanded—updated at best—to recognize the extreme threats to the individual or "self" that have arisen

---

[2]Russell L. Ackoff, et al., *Scientific Method* (New York: John Wiley & Sons, 1962).
[3]Russell L. Ackoff, *The Design of Social Research* (Chicago: University of Chicago Press, 1953).

as a result of modern technology and organizations themselves.[4] The Ideal of beauty also must be altered to recognize the ominous crossover between epistemology and aesthetics in the form of "unreality" that has occurred in recent years.[5] We also need to realize that every Ideal is part of a larger system of Ideals—no single Ideal exists or has meaning by itself.

The reason for our reformulation lies in the recognition of four major threats: (1) the threat to the whole of nature posed by modern technology; (2) the threat to the constitution and existence of the individual in Western societies; (3) the threat to the whole of human habitat or the living landscape; and (4) threats to the whole or basic constitution of knowledge itself. The first threat, the threat to the whole ecological system, arises from the fact that, as we have noted previously, for the first time in human history, human-induced crises have the potential to rival natural disasters in their scope and magnitude. Crises such as Bhopal, Chernobyl, and Exxon Valdez are unfortunately not rare, random aberrations. They are literally built into the very fiber and fabric of modern civilization. They are "normal" in that they are the natural and direct consequences of the kind of world we have created.[6]

Threats to the individual or the "self" arise from the existence of a class of organizations that can only be termed crisis prone.[7] Crisis-prone organizations not only constitute a major threat to the environment, but also to those who live and work in them. These organizations play and perpetuate a set of destructive "games" that Mitroff and Pauchant[8] have identified that literally threaten the physical and mental well being of their workers. For instance, one "game" they identified is "The Razor's Edge." Its objective is to "skate" as close to disaster as possible—for example, operating dangerous equipment without proper safety devices. The Razor's Edge is played in organizations where the normal work is devoid of challenge and workers are treated as subhumans.

We have unleashed on the physical landscape of America, and increasingly on the entire world, a series of architectural blights known as super or mega malls that embody the most grotesque mixtures of commercialism and entertainment. These constitute threats to the habitat. Western societies have succeeded beyond their wildest dreams in transforming large numbers of people into "unthinking, consumerist zom-

[4]Ian I. Mitroff and Thierry Pauchant, *We're So Big And Powerful Nothing Bad Can Happen To Us* (New York: Birch Lane Press, 1990).

[5]Ian I. Mitroff and Warren Bennis, *The Unreality Industry; the Deliberate Manufacturing of Falsehood and What It Is Doing to Our Lives* (New York: Birch Lane Press, 1989).

[6]Charles Perrow, *Normal Accidents* (New York: Basic Books, 1984).

[7]Mitroff and Pauchant, *op. cit.*

[8]*Ibid.*

bies."[9] This has been accomplished through the inducement of anxiety which can apparently only be relieved through the constant purchase of, or stimulation by, an endless variety of services and goods that are not necessary for human development. Especially in capitalist societies, people have literally been reduced to "consumption commodities." They are weighed, measured, and conceived of as "bundles of purchases." In other words, we are what we buy, or can be made to want.

One of the reasons why celebrities are so revered in Western societies is that they live out the ultimate fantasies of consumers. Presumably, they can purchase virtually anything, gain entrance to the best parties, shops, restaurants, and so on.[10] Essentially, however, they do not necessarily further human development conceived of in the broadest sense, that is, becoming the best person one is capable of.

The fourth threat involves the nature of knowledge itself, which has been radically and fundamentally transformed—"corrupted" is a far better word—by TV and its incorporation into a larger phenomenon that Mitroff and Bennis and Mitroff and Pauchant have labeled unreality.[11] As such they are following in the tradition of Umberto Eco[12] and Jean Baudrillard[13] who have termed the phenomenon hyper-reality and simulacra.

Television is so powerful a force that it has literally become the educational system of the United States. We have so thoroughly merged symbols of supposed "information," entertainment, and imagery that few can distinguish between them any longer. The line between reality, artificial reality, and unreality is now so thin it may have vanished altogether. For instance, a recent survey conducted by the Times-Mirror Corporation found that up to 50% of those who watch so-called "reality re-creation" crime shows such as "911" felt, when polled, that they were watching the "real thing." This was the case even though at the bottom of the TV screen clearly in front of each viewer was a distinct label saying that the scene was a re-creation or simulation. In short, we have created a whole new genre that we call *aesthetic epistemology* or, better yet, <u>*unaesthetic epistemology*</u>.

Western societies may have entered a new dark age. In contrast to earlier ones, this one has not been created by the suppression of knowledge, but by its prodigious production and unrelenting 24-hours-a-day

[9]*Ibid.*

[10]Stewart Ewen, *All Consuming Images, The Politics of Style in Contemporary Culture* (New York: Basic Books, 1988).

[11]Mitroff and Pauchant and Mitroff and Bennis, *op. cit.*

[12]Umberto Eco, *Travels in Hyper Reality* (New York: Harcourt Brace Jovanovich, 1986).

[13]Jean Baudrillard, *America*, translated by Chris Turner (London: Verso, 1988).

365-days-a-year total assault on our senses. In modern societies, citizens are bombarded by an overwhelming volume of disconnected "information" that mixes the banal, the entertaining, and the earth shattering in no order so that no one can make sense of it. This of course assumes that anyone still cares to make sense of it all.

We need much more than a mere restatement of Ideals to correct the preceding list of "evils." This is precisely why their correction is itself an Ideal. The four Ideals we envision are: (1) humane, whole person, and not "perfect" or its equivalent "masculine" knowledge; (2) replenishment and restoration of the environment, and not economic abundance or plenty; (3) goodness conceived of as a restoration of the "self," the dignity of the individual person, and the almost total redesign of our organizations so that they are not destructive to the environment; (4) aesthetic, not unaesthetic epistemology. At a minimum, we need to recognize that aesthetics and epistemology are much more intimately connected than we have thus far.

Consider the notion of humane knowledge. Some of the most interesting and we believe important recent developments in philosophy are to be found in the realm of feminist epistemology.[14]

Feminist philosophers are onto something important. They are profoundly and deeply skeptical of the foundations of Western philosophy because to a large extent they are correct in pointing out that it has largely been a masculine based, if not biased, endeavor. As a result, they believe that the basic ontology and epistemology of Western societies must be reoriented, if not reconstructed altogether. To give an example, it is enough to point out that in the West the primary, basic building blocks of reality (ontology) and knowledge (epistemology) have either been abstract properties of objects or states of mind out of which supposedly Truth arises. Feminist philosophers point out that this is so far removed from human experience that it constitutes a fundamental and serious distortion of how humans actually learn from the world and how they ought to relate to it.

All knowledge is relational, or in our terms "systemic," since it is grounded in how we develop as human beings. Our fundamental contact with the world is not initially with impersonal objects, but with that of a caring mother and father, and ideally, a caring community. Feminist philosophers are pushing for the creation of ontologies and epis-

[14]Ann Garry and Marilyn Pearsall (Eds.), *Women, Knowledge, and Reality, Explorations in Feminist Philosophy* (Boston: Unwin Hyman, 1989); Sandra Harding and Merrill B. Hintikka (Eds.), *Discovering Reality, Feminist Perspectives on Epistemology, Metaphysics, Methodology, and Philosophy of Science* (Dordrecht, Holland: D. Reidel, 1983).

temologies, or in our language ISs, that are based on ethical and caring relations between people, and not on dispassionate, impersonal knowledge of objects.

The consequences of this are profound and far reaching. If we are to reorient ourselves to the environment, if we are to transform crisis-prone organizations into crisis-prepared ones, then we need a different philosophy of humankind, of humankind's relationship to nature, of nature itself, of organizations—in short of an entirely different world. No one among us should say that Ideals do not have a fundamental role to play in this reconstruction.

The Ultimate Transformation: The World as an Integrated Whole

In the end, the greatest transformation is to see the world as an interconnected whole. Several recent accounts by contemporary philosophers offer testimony as to how far we are from this realization. For instance, Richard Rorty, who is hailed in some circles as "the most interesting philosopher in the world today," provides an important example.[15] His philosophy is decidedly nonsystemic.

Rorty notes that if contemporary philosophy has demonstrated anything, it is that there is no absolute perspective that enables us to say with absolute finality, "This is it!" Every picture, every model we construct of the world, is destined to be superseded by some other. Thus, for Rorty and for other contemporary philosophers, this establishes the downfall of all grand metaphysical systems, that is, systems that purport to explain all of reality by means of a limited number of general principles supposedly applicable in all times and all places. Instead, what is left is a world that is strongly contingent. Our lives are guided more by chance encounters and developments than we realize. The whole universe itself may be contingent. If so, literature may be a much better guide than philosophy in explaining this and helping human beings relate to a dynamically changing world.

(To say the least, this is Kant's problem taken to its ultimate extreme with a vengeance. The Executive, that is, the thing that picks the various views of a problem to present to a decision-maker, is perfectly arbitrary or random in the kind of world that Rorty describes.)

From the systems perspective we have been developing, Rorty's ar-

---

[15]Richard Rorty, *Contingency, Irony, Insolidarity* (Cambridge: Cambridge University Press, 1989).

gument fails miserably. His failure, however, is instructive for it reveals once again the importance of looking at problems and issues from the perspective of nonseparability, which is the heart of UST. Merely to *state*, to *contend*, that the world is contingent *is* to utter a general metaphysical principle. Rorty fails to ask the important questions: "What *kind* of world is it that permits pure or near pure contingency to operate?" and "What would the *structure* of the world have to be so that near pure contingency could operate?" Thus, whether he likes it or not, or whether he acknowledges it or not, Rorty makes an important metaphysical commitment. At the very least, he presupposes the kind of world in which strong contingency operates. But this is precisely a metaphysical statement, for such statements are beyond ordinary or scientific proof. One cannot inspect the whole universe to prove that it is contingent everywhere. From a systems perspective, Rorty's anti-metaphysics is thus a metaphysics nonetheless.

Next, while according to Rorty no picture of the world has priority over any other, Rorty's apparently does.

Most important, Rorty sees a fundamental split between metaphysics and aesthetics. From the standpoint of UST, this is not warranted. One can talk sensibly of aesthetic epistemology and even of aesthetic metaphysics—the two are not necessarily opposed. For instance, one can ask what kind of a deity or god might have created a world such that, while it is not absolutely contingent, it is "contingent enough" to permit individual actions to make a difference. Such questions do not prove that such a concept of God exists. Rather, through the pondering of such questions, we clarify our Ideals. This is perhaps one of their most fundamental purposes.

William James once posited that God's powers were limited so that humankind was "free" to develop and exercise moral choice. In James's view, the world was not complete. Although he didn't say it directly, his implication was that the lack of a final, complete picture of the world is the price we pay for human freedom. Humankind's actions therefore made a real difference in the outcome of the world:

[Suppose the author of the world put this case to you at the very moment of creation.] I'm going to make a world not certain to be saved, a world the perfection of which shall be conditioned merely, the condition that each . . . agent does its own "level best." I offer you the chance of taking part in such a world. Its safety, you see, is [not certain]. It is a real adventure, with real danger, yet it may win through. It is a social scheme of cooperative work generally to be done. Will you join the procession?

Will you trust yourself and trust the other agents enough to face the risks?[16]

The point is certainly not that this is a proof that James's version of God exists, for there are no such proofs. Instead, James's prose is both aesthetic and metaphysical at the same time, proof at least that the categories are not mutually exclusive or separable as Rorty assumes.

"God" is one of the greatest vehicles by which humans express Ideals and make them live. The concept of God is one of humankind's most inspired aesthetic and moral creations. In this sense at least, issues and matters of theology enter fundamentally into UST.

If one major contribution of the supposed "great philosophic minds" of this century is the assertion that there are no "final pictures of the world," then it pales in comparison to the great minds of earlier centuries. Kant is far more profound. We can anticipate what he might say in response to twentieth-century philosophers were he alive today.

Consider the earlier wine glass metaphor as a model for the human mind that we employed in Chapter 4. Because an infinite number of shapes are possible for wine glasses, this establishes that the assertion "There is no final, absolute shape for glasses" is true but in a trivial sense. The assertion misses the essential point that the property of "containment" is fundamentally critical to the concept of "glass," not that of any specific shape. Containment is in fact the most essential property. It is of little consequence that the particular form of every wine glass is variable. The most generic property that by definition is common to the concept of all wine glasses is what is important.

In a similar manner, the fact that there is no single, perfect, final language or model to capture the world misses the essential point that there is after all a language for expressing this very fact. The wondrous thing is not that we are unable to have perfect knowledge, but that we are able to have any knowledge at all. Even more wondrous is that we are able to learn at all. There must be some structure in the world that makes learning possible so that we can then write about it. What kind of world is it that is variable enough to permit human freedom, creativity, learning, and also incompleteness at the same time? This is the important question.

Since there is no final picture of the world that is open to human beings, this establishes that the decision to pick a particular picture on which to base one's actions is ultimately a heroic act and not a "logical"

---

[16]Patrick K. Dolley, *Pragmatism as Humanism, The Philosophy of William James* (Chicago: Nelson-Hall, 1974), pp. 155–156.

one. The choice of a particular action and the associated belief in it are among the greatest risks humans ever face.

Final Remarks

One of the most beautiful and moving accounts of interconnectedness was uttered by Chief Seattle in 1854. He gave a magnificent speech to an assembly of tribes who were preparing to sign a treaty with the White Man. He delivered his testimony in his native language, Duwamish. It was recorded by Dr. Smith who maintained afterwards that his translation did not do justice to the full beauty of the chief's imagery and thought:

> How can you buy or sell the sky, the warmth of the land? The idea is strange to us. If we do not own the freshness of the air and the sparkle of the water, how can you buy them? *Every part of this earth is sacred to my people* [italics ours]. Every shining pine needle, every sandy shore, every mist in the dark woods, every clearing and humming insect is wholly in the memory and experience of my people. The sap which courses through the trees carries the memories of the red man.
>
> The White Men instead forget the country of their birth when they go to walk among the stars. Our dead never forget this beautiful earth, for it is the mother of the Red Man. *We are part of the earth and it is part of us* [italics ours]. The perfumed flowers are our sisters; the deer, the horse, the great eagle, these are our brothers. The rocky crest, the juices of the meadows, the body heat of the pony, and man—*all belong to the same family* [italics ours] . . .
>
> The ashes of our father are scared. Their graves are holy ground, and so these hills, these trees, this portion of earth is consecrated to us. We know that the white man does not understand our ways. One portion of land is the same to him as the next, for he is a stranger who comes in the night and takes from the land whatever he needs. The earth is not his brother, but his enemy, and when he has conquered it, he moves on. He leaves his fathers' graves behind him, and he does not care. He kidnaps the earth from his children. He does not care. His fathers' graves and his children's birth right are forgotten. He treats his mother, the earth, and his brother, the sky, as things to be bought, plundered, sold like sheep or bright beads. His appetite will devour the earth and leave behind only a desert. . . .
>
> I am a savage and I do not understand any other way. I have seen a thousand rotting buffaloes on the prairie, left by the white man who shot them from a passing train. I am a savage and I do not understand how the smoking iron horse can be more important than the buffalo that we kill only to stay alive. . . .

What is man without the beast? If all the beasts were gone, men would die from great loneliness of spirit. *For whatever happens to the beast, soon happens to man. All things are interconnected* [italics ours]. . . .

You must teach your children that the ground beneath their feet is the ashes of our grandfathers. So that they will respect the land, tell your children that the earth is rich with the lives of our kin. Teach your children what we have taught our children, that the earth is our mother. *Whatever befalls the earth befalls the sons of the earth. If men spit upon the ground, they spit upon themselves* [italics ours]. . . .

Where is the thicket? Gone. Where is the eagle? Gone. And what is it to say goodbye to the swift pony and the hunt? The end of living and the beginning of survival.

This we know. The earth does not belong to man; man belongs to the earth. This we know. *All things are connected like the blood which unites one family. All things are connected* [italics ours].

Whatever befalls the earth befalls the sons of the earth. Man did not weave the web of life, he is merely a strand in it. Whatever he does to the web, he does to himself [italics ours]. . . .

The whites too shall pass; perhaps sooner than all other tribes. Continue to contaminate your bed, and you will one night suffocate in your own waste. . . .

When the last red man has vanished from this earth, and his memory is only the shadow of a cloud moving across the prairie, these shores and forests will still hold the spirits of my people. For they love this earth as the newborn loves its mother's heartbeat. So if we sell you our land, love it as we've loved it. Care for it as we've cared for it. Hold in your mind the memory of the land as it is when you take it. And with all your strength, with all your mind, with all your heart, preserve it for your children and love it.[17]

---

[17]David Suzuki, *Inventing the Future: Reflections on Science, Technology, and Nature* (Stoddart, Toronto: 1989), pp. 230–231.

APPENDIX

Deeper Aspects of Unbounded Systems Thinking

At the heart of the "sweeping in" process that is at the core of UST is a theory of how all the branches of knowledge relate to one another.

Two archetypal patterns emerge repeatedly in exploring the relationship between the various sciences, branches of knowledge, and ISs we have introduced. One is a strict hierarchical pattern; the other, a complete lack of hierarchy. In the first pattern some—typically one or a select few—sciences are regarded as the most basic or fundamental; in the second, all the sciences and all the various branches of knowledge are on an equal footing. The first pattern is inherently reductionistic; the second, inherently systemic.

Nowhere are these two archetypes presented and contrasted more vividly than in the works of C. West Churchman[1] and his mentor, E. A. Singer.[2] Churchman and Singer depict the strongest possible form of the hierarchical archetype.

In doing so, their purpose is not to argue for it, but rather to critique it, and show that their own philosophy of inquiry is founded on an opposite archetype.

Churchman and Singer start with the following questions: What if the sciences were organized in a linear, hierarchical fashion? What if the sciences were arrayed from the supposedly most basic or most fundamental down to the least fundamental or least basic? What science or sciences would head the list? What would the relation between the sciences be? The work of the positivists supplies a response to all these questions.

---

[1]C. West Churchman, *The Design of Inquiring Systems* (New York: Basic Books, 1971).

[2]E. A. Singer, *Experience and Reflection*, C. West Churchman (Ed.) (Philadelphia: University of Pennsylvania Press, 1959).

Logic and mathematics have repeatedly been taken as the most fundamental of all the sciences. This is because, in its basic desire to reason clearly, every science presupposes the prior existence of the science of logic. All sciences place an extremely high premium on the concept of logical consistency. It is simply taken for granted that a scientific assertion or proposition cannot be both true and false at the same time. Thus logic, even more than mathematics, is the leading candidate for the most basic of all the sciences. Further, the relationship between the sciences is established by means of the concept of *presupposes*. Thus, science X is more basic than science Y, if Y presupposes (i.e., makes use of) the concepts of science X, but X does not presuppose the concepts of Y.

Having introduced the most basic science—logic—and the relationship between the sciences—the concept *"presupposes"*—Churchman and Singer illustrate the resulting ordered list of sciences. We will not pursue in detail the complete listing of the sciences below logic. For our purposes, it is more important to note that the taxonomy of the sciences is split in half. In the top half are arrayed the mathematical and the physical sciences, with logic at the top. In the bottom half are the social sciences. This reflects the common notion of the supposed superiority of the physical sciences over the social in every conceivable dimension. Supposedly they are more exact, more precise, more theoretically sound, and so on.

Two other important ideas lurk within the hierarchy. Every science introduces special terms, concepts, and ideas. For instance, logic introduces the concept of "logical laws," logical operators, elementary propositions, sentences, valid forms of reasoning, incorrect forms of reasoning, and so on. Physics introduces concepts such as energy, force, mass, spatial coordinate systems, and conservation laws of energy.

A crucial issue in all such orderings is whether the concepts of science Y farther down the list can be strictly derived from or assimilated in terms of the concepts of science X higher up on the list. If all the concepts of science Y can be shown to follow from those of X, then science Y can be said to be *reducible* to that of X. This is the thesis of *reductionism.*

The concept of *presupposes* is looser than that of reductionism. To apply the concept of "presupposes," we need merely show that *some* of the more important aspects of science Y depend on those of science X. We do not have to demonstrate that *all* of Y can be reduced to or is contained within all of X. Reductionism is thus the strongest possible way of ordering the list of the various sciences.

A critical juncture is reached when we encounter the science of biology. As members of the physical world, all living things must be subject to physical and chemical processes and laws. Although this is true, it raises an important difficulty. We are not yet able to reduce all of complex, human, mental, and social phenomena to, say, neurophysiological models and laws—although or capability has grown enormously in this area. The difficulty with explanations of living things solely or purely in physical terms is that, for every physical law or rule that purports to separate clearly the living from the nonliving, one can always find certain nonliving things that somehow pass the rule and ought therefore to count as living things, and certain living things that fail the rule and ought therefore to be categorized as nonliving. It is not that there are no differences between the two classes—there obviously are—but rather that many of the differences are extremely difficult, if not impossible, to capture in physical laws or mathematical formulas. For example, certain insects neither reproduce themselves nor display metabolic processes, while certain machines do "reproduce," and certain chemicals, and even buildings display "metabolic" processes. For instance, modern buildings are now designed to turn themselves on and shut themselves down in response to various light and chemical changes over the course of a 24-hour period to save energy.

Modern philosophers of biology have been forced time and time again back to the concept of teleology or purposiveness in order to separate the class of living from nonliving things.[3] The concept of living things is not completely reducible to physical laws, processes, or mathematical formulas.[4]

Biology has brought us to the lower half of the taxonomy, the social sciences. Or, more accurately, biology straddles the fence between the physical and the social sciences. The various social sciences can be distinguished by the size and complexity of the social organisms they treat. Thus, if biology entails some of the smallest and lowest orders of living things, then psychology is the first of the social sciences to treat higher-order living things. In particular, the newer humanistic and psychoanalytic psychologies purport to treat "the whole human being," and not merely one or more of its special parts or aspects.

If psychology generally treats the autonomous, self-contained, whole human being, then the subject matter of sociology ranges from that of

---

[3]C. West Churchman, *The Design of Inquiring Systems* (New York: Basic Books, 1971); and E. A. Singer, *Experience and Reflection*, C. West Churchman (Ed.) (Philadelphia: University of Pennsylvania Press, 1959).

[4]By "teleology" we do not mean a mysterious "elan vital" that inhabits all living creatures, but rather the scientific study of purposes that are imputed by biological scientists to the organisms they study.

small groups to whole societies. It is here that reductionism raises its head again. If a group is no more than a sum of its parts, then the science of sociology in general—and at the very least the subject matter of small-group behavior—ought to be reducible to psychology or psychological phenomena. The crucial word is "if." The important question is whether all group behavior is reducible to individual behavior, or whether groups have some properties that no one of its individual members possess. And indeed there are. For instance, national defense is a property of society taken as a whole, not that of any one of its individual parts.

If sociology uses as its unit of analysis the concept of the group and society, then anthropology takes the concept of culture as its fundamental precept and focuses on the systematic differences between whole societies. At the same time, we must realize that there are extreme overlaps between all the social sciences because there are no exact dividing lines. Of necessity, we can only discuss the strongest differences between the various sciences here.

History, on the other hand, treats the largest of all human groups—humankind—and seeks to discover whatever generalized patterns of actions and purposes characterize human societies viewed in the broadest possible sense. History has only recently, and with considerable reluctance, been admitted as a social science. Thus the hesitation to consider theology as a social science is even more understandable. Leaving aside this debate, we can nonetheless recognize that theology treats the largest human experience imaginable—contemplation of the nature of the entire universe.

This completes the linear, hierarchical archetype of the ordering of the sciences. It is linearly hierarchical because it views the sciences as proceeding strictly from the most theoretically certain and rigorous down to the least theoretically certain and rigorous. There is, however, another relationship between the sciences that opposes the linear archetype in every respect. This opposite archetype can be called a circle or a wheel to distinguish it from the linear one. The wheel archetype is less commonly recognized because it is based on a less widely practiced style of inquiry. It goes against the grain of traditional scientific thinking and the premises that underlie the current design of the modern university, as well as most institutions in Western societies.

The circular archetype is founded on the radical notion that *all the sciences presuppose one another* in the sense that *the concepts of all the sciences have a bearing on one another*. From this perspective, there are no absolutely fundamental or more basic sciences and professions. Some sciences may be more highly developed along some criteria than others, but there is no science that is equally well developed along all lines or criteria. Thus, if one science is more basic according to some criteria, it will invariably be less basic according to others. The wheel archetype regards the relationship between the various sciences as one of mutual complementarity and support, not one of opposition or competition.

Consider, for example, physics and sociology. There is no doubt that, historically, sociology owes a tremendous debt to physics. Physics has continually supplied sociology with a powerful set of organizing concepts and metaphors. For example, some of the earliest attempts to build explanatory models of society were clearly governed by Newtonian thinking. By analogy, society was conceived as a mass acted on by a series of propelling and restraining forces. The net sum of these forces caused society to move in certain directions with a certain acceleration, and so forth. Thus, society no less than physical phenomena were subject to the Newtonian law that force equals mass times acceleration. What is less clear is that sociology has anything to offer so exalted a science as physics. Some, such as Karl Popper,[5] deny this vehemently. However, it is precisely this proposition that the wheel archetype or UST asserts. The wheel archetype contends that the science of physics has presupposed, without its conscious awareness, a complex social organization. It is precisely because physics has taken its social side for granted that it has failed to acknowledge its debt to social sciences such as sociology.[6]

(In a more critical vein, it is appropriate to note that the social sciences are burdened by the long baggage of the history of the physical sciences. If the physical sciences had not developed first and had not been so successful, then perhaps the social sciences might have been freer to develop in ways that were much more appropriate to their subject matter.)

More specifically, all science is the product of a complex set of social and institutional forces that have brought it into being and sustained its existence. For all practical purposes, physics today is the leading member of Big Science. We are far from the days of the great physicist Maxwell playing around with magnets in his garage. Today's physicists play instead with very big and costly machines, cyclotrons and bevatrons, which are used to penetrate the structure of the atom. As a result, they demand a complex social and institutional structure both to design and operate them. For instance, not every physicist who so desires can gain

[5]Karl Popper, "Normal Science and Its Danger," in Imre Lakatos and Alan Musgrave (Eds.), *Criticism and the Growth of Knowledge* (New York: Cambridge University Press, 1970).
[6]Ian I. Mitroff, *The Subjective Side of Science* (Amsterdam: Elsevier, 1974).

access to these machines in order to perform his or her favorite experiment.

One could contend that the criteria used to allocate precious time on atom smashers, a valuable and costly resource, are purely objective and flow perfectly from the principles of physics so that physics has no need of sociology or modern management principles. Unfortunately, however, growing evidence—and scientific evidence at that—from psychology and sociology does not support these contentions.[7] To the contrary, the evidence is that a wide variety of criteria enter into the selection of those who are allowed to run their experiments, and that many of these criteria are social—for instance, the experimenter's perceived standing in the physics community, the prestige of the university where he or she is currently located, and so on. It's not that this is necessarily wrong. Such criteria operate in all social systems. Our physical science would have to be far more developed than it may ever be to have a set of purely objective rules for selecting the best experiment out of all those proposed. If anything is wrong, it is the lack of *conscious acknowledgment* of the tremendous role that social processes actually play in physical science.

Let us examine another example. Consider the fact that nuclear physics often demands the scanning of literally thousands of photographs in order to detect the occurrence of a significant, but rare, nuclear event as evidence for the existence of a new subatomic particle. Whether the scanning operation is completely automated or not, humans have to be involved at some point in the process to design the automation procedures and verify the results derived from them. Thus, a psychological entity of some kind is presupposed. Further, it is no easy or trivial task to motivate individuals to maintain a high level of awareness to select out rare events from thousands of background events. The design of an effective monitoring system is not entirely a physical-science problem. If one pursues the issue entirely in physical terms, one neglects important knowledge and variables from another science. It's like looking for a quantum mechanical explanation of love when we suspect that psychological, literary, or philosophical ones might just be a bit more fruitful.

For another, consider that in engineering we find that technological risks cannot be understood purely in technical terms such as equipment mean time to failure and probability analyses. As the Kemeny Commission on the Three-Mile Island nuclear accident concluded:

[7]*Ibid.*

> The fundamental problems are people-related problems and not equipment problems. . . . [W]herever we look, we found problems with the human beings who operate the plant, with the management that runs the key organization, and with the agency that is charged with assuring the safety of nuclear power plants.[8]

From the standpoint of the applied social sciences, the linear archetype is not just inappropriate, but downright silly—an outworn image of an earlier age. For this reason, a fundamental tension between the so-called pure and the applied sciences will probably always exist. The pure sciences generally adhere to the linear archetype. To use Alvin Toffler's terminology, during the Third-Wave, the electronic age, in which everything affects everything else, this world view is no longer warranted. We can no longer afford to separate the various sciences or rely on the outmoded philosophies of positivism and reductionism.

In a beautifully written but sadly neglected book, *Reason in Society*, Paul Diesing[9] illustrates the extreme interconnectedness we have been discussing. Diesing compares and contrasts explicitly and systematically four different concepts of rationality and shows, to the dismay of those who have proposed their separate and primary existence, that each is completely dependent on the other. The four modes of rationality are: economic, legal, political, and social.

Despite all the complex models of economists and economic arguments, a basic idea lies at the core of economic reasoning: the presumption of a clear "economic yardstick" or a "basic economic unit of analysis" in terms of which all things of value may be measured or weighed. Legal rationality sets out a framework for settling basic disputes in society. Political rationality specifies a framework for citizen representation and participation in the affairs of state. Social rationality specifies a series of legitimate distinctions among humans, for example, the unborn, so that society can govern itself.

Diesing shows with uncommon clarity, rigor, and vigor that each of these modes of rationality presupposes one another. For example, there could be no such thing as a "basic economic yardstick" unless there was already a preexisting, stable society, a well-accepted legal framework, and a series of accepted social strata. So it is with all the other concepts of rationality. Each presupposes the other in the sense that neither of them could exist or operate without the others.

It is always deliciously tempting to take every concept out of context,

---

[8]J. Kemeny, et al., *Report of the President's Commission on the Accident at Three Mile Island* (New York: Pergamon Press, 1979), p. 8.
[9]Paul Diesing, *Reason in Society* (Champaign: University of Illinois Press, 1971).

and, by thus severing its relationship to that of others, to perceive each as existing separately and independently. But as Diesing shows so cogently, it is not possible to do this. *Implicit in the very concept of each notion of rationality is the implied necessity of all the others.* The necessity of each concept becomes apparent as soon as we analyze each one in detail.

Our survey of Singer and Churchman can be summarized in a few brief but critical propositions:

- Every science is to be found within every other.
- Every model presupposes every other model.
- Every problem is to be found within every other problem.

The recognition of and use of these insights or principles is precisely what characterizes or constitutes UST. As such, UST is broader than the Multiple Perspective Method described earlier. In addition to sweeping in every known discipline or profession as a *potential resource* to illuminate every problem, UST also insists on including a broader sense and scope of ethics and aesthetics as two of the most vital aspects of every problem. What we call a "problem" is not only a reflection of our values but of our ethical commitments, of what we believe ought to be. Especially in the social realm, something is a problem because things are not as they ought to be. Thus, the gaps between what we desire and what we can accomplish are not merely measured by T, O, and P. Instead, they constitute ethical and aesthetic gaps as well. Consideration of aesthetics and ethics thus play a fundamental role in our selection of problems and in the means we use to address them.

References

Abegglen, J. C., and Stalk, G., Jr. *Kaisha, The Japanese Corporation* (New York: Basic Books, 1985).

Abernathy, W. J., et al. *Industrial Renaissance: Producing a Competitive Future for America* (New York: Basic Books, 1983).

Ackoff, R. L., and Emery, F. E. *On Purposeful Systems* (New York: Aldine-Atherton, 1972).

Bluestone, B., and Harrison, B. *The Deindustrialization of America* (New York: Basic Books, 1982).

Churchman, C. W. *The Design of Inquiring Systems* (New York: Basic Books, 1971).

Fallows, J. "America's Changing Economic Landscape," *Atlantic* (March, 1985), p. 56.

Gevirtz, D. *Business Plan for America: An Entrepreneur's Manifesto* (New York: Putnam, 1984).

Linstone, H. A. *Multiple Perspectives for Decision Making, Bridging the Gap Between Analysis and Action* (New York: North-Holland, 1984).

Mitroff, I. I. *The Subjective Side of Science: A Philosophical Inquiry into the Psychology of the Apollo Moon Scientists* (New York: Elsevier, 1974).

Phillips, K. *Staying On Top: The Business Case for a National Industrial Strategy* (New York: Random House, 1984).

Piore, M. J., and Sabel, C. F. *The Second Industrial Divide* (New York: Basic Books, 1984).

Reich, R. B. *The Next American Frontier* (New York: Times Books, 1983).

Reich, R. B. *Tales of a New America* (New York: Times Books, 1987).

Thurow, L. *The Zero-Sum Solution: Building a World-Class American Economy* (New York: Simon and Schuster, 1985).

Index